T0227679

SOIL
BIOCHEMISTRY

SOIL BIOCHEMISTRY

Konrad Haider
Augustinum, Diessen, Ammersee
Germany

and

Andreas Schäffer
Institute for Environmental Research (Biology V)
RWTH Aachen University
Germany

CRC Press
Taylor & Francis Group
Boca Raton London New York

CRC Press is an imprint of the
Taylor & Francis Group, an **informa** business

Science Publishers
Enfield (NH)　　Jersey　　Plymouth

CRC Press
Taylor & Francis Group
6000 Broken Sound Parkway NW, Suite 300
Boca Raton, FL 33487-2742

First issued in paperback 2017

© 2009 by Taylor & Francis Group, LLC
CRC Press is an imprint of Taylor & Francis Group, an Informa business

No claim to original U.S. Government works

ISBN-13: 978-1-57808-579-8 (hbk)
ISBN-13: 978-1-138-11630-6 (pbk)

This book contains information obtained from authentic and highly regarded sources. While all reasonable efforts have been made to publish reliable data and information, neither the author[s] nor the publisher can accept any legal responsibility or liability for any errors or omissions that may be made. The publishers wish to make clear that any views or opinions expressed in this book by individual editors, authors or contributors are personal to them and do not necessarily reflect the views / opinions of the publishers. The information or guidance contained in this book is intended for use by medical, scientific or health-care professionals and is provided strictly as a supplement to the medical or other professional's own judgement, their knowledge of the patient's medical history, relevant manufacturer's instructions and the appropriate best practice guidelines. Because of the rapid advances in medical science, any information or advice on dosages, procedures or diagnoses should be independently verified. The reader is strongly urged to consult the relevant national drug formulary and the drug companies' and device or material manufacturers' printed instructions, and their websites, before administering or utilizing any of the drugs, devices or materials mentioned in this book. This book does not indicate whether a particular treatment is appropriate or suitable for a particular individual. Ultimately it is the sole responsibility of the medical professional to make his or her own professional judgements, so as to advise and treat patients appropriately. The authors and publishers have also attempted to trace the copyright holders of all material reproduced in this publication and apologize to copyright holders if permission to publish in this form has not been obtained. If any copyright material has not been acknowledged please write and let us know so we may rectify in any future reprint.

Except as permitted under U.S. Copyright Law, no part of this book may be reprinted, reproduced, transmitted, or utilized in any form by any electronic, mechanical, or other means, now known or hereafter invented, including photocopying, microfilming, and recording, or in any information storage or retrieval system, without written permission from the publishers.

For permission to photocopy or use material electronically from this work, please access www.copyright.com (http://www.copyright.com/) or contact the Copyright Clearance Center, Inc. (CCC), 222 Rosewood Drive, Danvers, MA 01923, 978-750-8400. CCC is a not-for-profit organization that provides licenses and registration for a variety of users. For organizations that have been granted a photocopy license by the CCC, a separate system of payment has been arranged.

Trademark Notice: Product or corporate names may be trademarks or registered trademarks, and are used only for identification and explanation without intent to infringe.

Library of Congress Cataloging-in-Publication Data

Haider, Konrad, 1928-
[Biochemie des Bodens. English]
Soil biochemistry / Konrad Haider and Andreas Schäffer.
    p. cm.
Includes bibliographical references and index.
ISBN 978-1-57808-579-8 (hardcover)
1. Soil biochemistry. I. Schäffer, Andreas. II. Title.
S592.7.H3513 2009
631.4'17—dc22

                                                                    2009016941

Visit the Taylor & Francis Web site at
http://www.taylorandfrancis.com

and the CRC Press Web site at
http://www.crcpress.com

Preface

This is the improved and enlarged English version of a German book, which appeared in 1996 as "Biochemie des Bodens" published by Ferdinand Enke Verlag, Stuttgart. For preparing this new edition of the book we are indebted to Prof. Dr. Jean-Marc Bollag, Center for Bioremediation and Detoxification Research Institute, The Pennsylvania State University, University Park, Pensylvania, for his very valuable comments for improving the manuscript. Our thanks goes equally to one of his editorial coworkers, Dr. Joy Drohanr, for her corrections as well as to many friends and colleagues, who were of great help in submitting data and figures used in the text.

Konrad Haider and Andreas Schäffer

Preface

Contents

List of Figures

List of Tables

Soil and Soil Life

Soils contain a lively biota consisting of a great number of bacteria, fungi and animal species. One cm^3 of fertile soil may contain up to 10^{10} bacteria belonging to an estimated 10^4 species. Fungi are more difficult to enumerate, but under slightly acid soil conditions as in forest soils their biomass surpasses that of the bacterial biomass (Anderson and Weigel, 2003). Fungi are the most important soil microorganism group involved in the carbon cycle, with 40–200 g of mycelial dry matter per square meter of soil (Fig. 1.1).

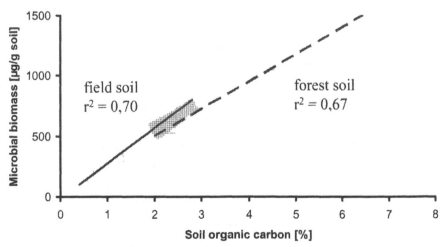

Fig. 1.1: The respiratory response of fungi and bacteria in relation to soil pH (Anderson and Weigel, 2003).

A survey of the average contribution of fungi and bacteria to soil microbial biomass by combining microscopy, selective inhibition, ergosterol content, and total microbial tissue by fungal glucosamine and bacterial muramic acid shows a fungal contribution of between 30 and 85% and a bacteria contribution between 20 and 30% (Anderson and Weigel, 2003).

Fungal extracted DNA yields from fungi are lower than that for bacteria and amounts by 5 μg mg^{-1} dry mycelia weight (Anderson, 2008).

Based on their nutritional mode fungi can be divided into saprophytes, symbionts, and parasites. Despite their great numbers they occupy only a small area of the surfaces available in soils (micropores) and they prefer to settle in pores with diameters of 0.2 to 10 μm where they are protected against the attacks of protozoa.

Soil pore size influences distribution of soil biota

Soil bacteria, which typically average 0.2 to 1.0 μm in size reside preferentially in pores ranging from 2.5–5 μm in diameter for fine- and coarse-textured soils, respectively (Fig. 1.2). Few bacteria have been observed to reside in pores less than 0.8 μm in diameter. This means that 20 to 50% of the total soil pore volume cannot be accessed and utilized by the microbial community. Bacterial cells are often embedded in mucilage, a sticky polysaccharide material of bacterial origin that is attached by clay particles providing protection against desiccation, predation, and noxious compounds.

Fungi, protozoa, and algae are mainly found in pores larger than 5 μm. Fungi and their mycelia are commonly observed on surfaces and sometimes develop extensive networks that bind soil particles to aggregates.

The contribution of the various kinds of soil biota to the overall soil biomass is shown in Table 1.1

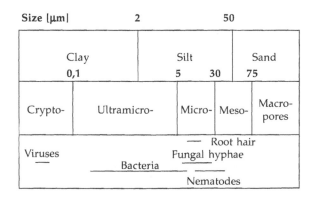

Fig. 1.2: Size of soil particles and soil biota in the USDA classification.

Table 1.1: Contribution of the various kinds of soil biota to soil biomass (adapted from Gisi et al., 1990).

Organisms	Biomass with plant roots %	Biomass without plant roots %
Subterrestrial components of plants	50	–
Fungi	29	58
Bacteria	13	25
Earthworms	5	10
Protozoa	2	4
Nematodes	0.3	1
Various	0.7	2

Soil water and aeration

Oxygen diffuses 10,000 times as fast through the air spaces as it does through water in soil. This restricts microorganisms. A minimum of 10% airfilled pore space is commonly considered necessary for adequate aeration. Clay soils with 45% water commonly have inadequate pore space for aeration if their bulk density is greater than 1.3 g/cm^3, and total pore space is less than 50%.

The change from aerobic to anaerobic metabolism occurs at oxygen concentrations of less than 1%. The fact that anaerobic processes such as denitrification (see 4.4) and sulfate reduction occur in many well aerated soils indicates that anaerobic microsites occur commonly in many soils. Further evidence that anaerobic microsites exist can be deduced from the common occurrence of anaerobic bacteria like Clostridia in the upper layers of soil (Paul and Clark, 1989, 1996).

Methods for the quantification of microbial biomass

The aim of this section is to provide a short survey of the main methodological tools that soil microbiologists and ecologists have at their disposal to study soil biomass quantities (Fig. 1.3).

Fumigation incubation (FI) and fumigation extraction (FE) methods

Methodologies for quantification of the microbial biomass by cell counting or cultivation are described and reviewed in "Methods in Soil Biology" by Schinner et al. (1996) as well as by Martens (1995).

The use of a variety of molecular techniques generally demonstrated that soil environments are far more diverse and contain previously undescribed groups as important components of the soil community (Tiedje et al., 1999).

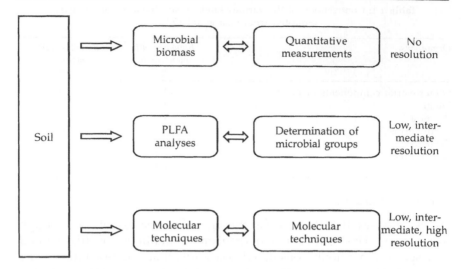

Fig. 1.3: Methods for quantification of microbial biomass.

Jenkinson and Powlson (1976) reported that extra CO_2 evolving from soil after fumigation with $CHCl_3$ came mainly from decomposition of microbial cells and that this extra flush could be used as a measure of the microbial biomass in soil. They said that the evolved CO_2 should be proportional to the quantity of biomass carbon (Bc) from the killed cells and developed the equation $Bc = Fc/k_c$ (Fig. 1.4) where Fc is the flux of CO_2-C during a 10 day incubation at 25 °C from the fumigated soil minus CO_2 evolved from the unfumigated soil. The k_c factor represents the portion of carbon from the killed biomass mineralized to CO_2; under standarized conditions and at a constant temperature of 25° C it can be set for a 10 day incubation to 0.45 (the unit of kc is (incubation time)$^{-1}$).

The k_c factor can not be used for acidic soils or for soils with a residue of incompletely decomposed plant residues, because this material is mineralized faster in a fumigated soil than in the nonfumigated soil (Martens, 1995).

Fumigation-extraction method for carbon, nitrogen, phosphorus and sulfur

The fumigation-extraction method (FE method) overcomes the limitations of the FI-method which is not to be applied to acidic soils or soils with recently added substrates. The FE method is based on the observation that fumigation causes an immediate increase in the amount of organic carbon that can be extracted by aqueous solutions. The amount of extracted C_{org} can be multiplied by the empirical factor of 2.64 for conversion into biomass carbon. Methods for measuring nitrogen, phosphorus, and sulfur

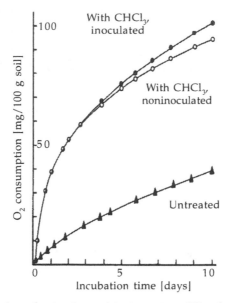

Fig. 1.4: Chloroform fumigation and its impact on CO_2 release (Haider, 1996, according to Jenkinson and Powlson, 1976).

in the extracts after fumigation were published by Brookes et al. (1982). Both, the FI and FE methods can be used to investigate the amount of ^{14}C or ^{15}N incorporated into the biomass from applied ^{14}C- or ^{15}N-labeled substrates (Stott et al., 1983).

Substrate-induced respiration (SIR) method

Anderson and Domsch (1978) proposed a physiological method for the measurement of soil microbial biomass based on soil respiration in response to the addition of a readily decomposable substrate such as glucose. The short period of constant respiration occurring for a few hours before the onset of rapidly increasing respiration due to microbial growth was interpreted as being proportional to the size of the biomass at the addition of glucose and was correlated to the flush of decomposition following $CHCl_3$ fumigation of the FI and FE methods (Fig. 1.5). On average, in moist soils tested 40 mg of microbial biomass mineralized the added glucose at an initial maximum rate of 1 mL CO_2/h.

Adenosine triphosphate (ATP) and adenylate energy charge-physiological activities of soil microbial populations

Jenkinson and Oades (1979) demonstrated a close linear relationship between ATP contents and biomass carbon as measured by the FI method.

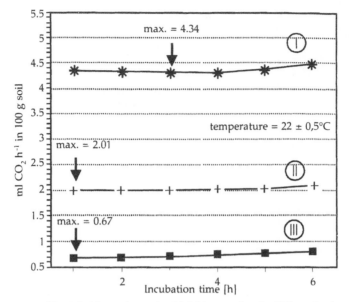

Fig. 1.5: Measuring microbial biomass by the SIR method.
(I), (II), and (III) refer to different glucose concentrations.

The solution used to extract ATP from soil is critical; strongly acidic reagents such as H_2SO_4 or trichloroacetic acid are recommended to deactivate ATPase within cells, whereas neutral reagents like $NaHCO_3$ or $CHCl_3$ are not sufficient (Jenkinson and Oades, 1979). Recovery of ATP by adding a ATP-spike can be corrected. Assuming that biomass contains 46% carbon, it can be calculated that 11.7 µmol of extracted ATP equalizes 1 g biomass.

Adenylate energy charge (AEC) is a measure of the metabolic energy stored in the nucleotide pool and is defined as AEC = ([ATP] + 0.5 [ADP])/ ([ATP] + [ADP] + [AMP]). Theoretically, AEC can range from 0 (all AMP at low energy) to 1.0 (all ATP at high energy). Values found by Brookes et al. (1983) were 0.85 for freshly sampled grassland soil to less than 0.1 for soils exhausted by consumption of nutrients (Martens, 1995).

Many changes in land use or agricultural management lead to changes in soil organic matter content. Because they occur slowly the changes are difficult to measure in the short- or medium-term against a large organic matter background. The quantity of microbial biomass in soil is a much more sensitive parameter that could be used as early warnings of eological changes.

Brookes (1993) suggested that biomass carbon expressed as a percentage of total soil organic carbon (C_{mic}/C_{org} ratio) could be used as an indicator of stress of microorganisms due to mismanagement or soil contamination (Alef et al., 1988). Beck (1990), as well as Anderson and Domsch (1989), showed that this ratio in soils with a regular crop rotation is higher than that in soils with long-term monocultures. From soil enzymatic activities together with microbial biomass contents, several research teams have tried to develop a significant signature indicating whether or not a certain method of soil management or soil stress due to contamination can lead to a long-term increase or decrease of soil organic matter contents and soil fertility (Beck,1984a, b; Powlson et al., 1987).

Extracellular enzymes in soil

Besides being present inside micoorganisms or plant residue particles, enzymes also exist extracellularly in the soil solution or are attached to clay surfaces or adsorbed to humic compounds (Dick and Kandeler, 2005; Beck and Beck, 2000). It is generally assumed that soil enzymes are of microbial origin, and that origin from animals or plants play only a minor role. There is, however, an increased enzymatic activity in rhizosphere soil because roots excrete extracellular enzymes and they can be excreted by sloughed off cellular debris or by elevated microbial activity in the root zone.

Approximately 100 enzymes have been identified in soils. The most prominent are oxidoreductases, transferases, and hydrolases (Table 1.2). The oxidoreductase dehydrogenase has been widely studied because of its apparent role in the oxidation of organic matter, where it transfers hydrogen from substrates to acceptors. Some hydrolases and transferases have been extensively studied because of their role in decomposition of various organic compounds and in enzymes involved in nutrient cycling and formation of soil organic matter. These include enzymes such as amylase, cellulase, xylanase, glucosidase, and invertase, and enzymes involved in the N cycle such as protease, amidase, urease, and deaminase, in the P cycle such as phosphatase, and the S cycle such as arylsulfatase.

Extraction of Enzymes from Soil

Several extraction procedures for different enzymes are described by Tabatabai and Fung (1992). There is, however, limited access to enzymes by extraction from soils because they are complexed with the soil matrix and lose their integrity during the extraction procedure. Clays and complexes of organic matter are known to bind proteins, generally slowing down the degradation by microorganisms or proteases. Their protein cannot be separated easily from mineral or organic colloids, or it is denatured during the extraction procedure. A large number of assays have been developed to be performed at substrate concentrations exceeding enzyme saturation on a known amount of soil and under conditions that include a certain range of temperature, pH, and ionic strength. Protocols for the assays generally

Table 1.2: Enzymes and enzyme classes detected in soil
(Tabatabai and Fung, 1992; Beck and Beck, 2000).

Enzyme classes	Name and reaction	Substrate
Oxidoreductases		
Catalase	$2\ H_2O_2 \rightarrow O_2 + 2\ H_2O$	H_2O_2
Peroxidase	Donor + $H_2O_2 \rightarrow$ oxidized donor	Pyrogallol, chloroanilines, dianisidine
Monooxygenase	Phenylalanine + $O_2 \rightarrow$ tyrosine	Phenylalanine
	Phenol or benzene + O_2	Catechol, *p*-cresol, phenyldiamin
Lyases		
Carboxylesterase	Carboxylester + $H_2O \rightarrow$ alcohol + carboxylic acid	
Arylesterase	Phenylacetate + $H_2O \rightarrow$ phenol + acetate	
Phosphatases	Phosphoricacid esters + $H_2O \rightarrow$ alcohol + orthophosphate	
Hydrolases		
Cellulase	Endohydrolysis of 1,4–β glycosidic linkages in cellulose	Cellulose, cellobiose
β-Glucosidases	Hydrolysis of terminal β-D-glucose residues with release of β-D-glucose	Cellobiose
Proteinases	Hydrolysis of proteins to peptides and amino acids	Casein, gelatin, albumin
Urease	Urea + $H_2O \rightarrow CO_2 + NH_3$	Urea
Aromatic-L-amino acid decarboxylase	L-tryptophan → tryptamine + CO_2	DL-3,4-dihydroxyphenyl-alanine, DL-tyrosine, DL-phenylalanine, tryptophan

include antiseptic conditions such as toluene to inhibit growth and metabolism during the assay. Many of these assays utilize spectrophotometric analysis of the reaction products.

The stability of many enzymes in soils has been demonstrated and shown to increase by complexation to soil humic material (e.g. Nannipieri et al., 2003). MacLaren (1975) showed that a urease-organic matter complex is more resistant to attack by a proteinase than the pure urease (Dick and Kandeler, 2005).

There has been great interest in developing soil enzyme assays as indicators of soil quality to reflect impacts of pollution or other alteration in soil quality. Field-scale studies have shown that enzymes are affected by soil management activities. In many cases they are early predictors of soil management effects due to farming practises such as crop rotation, e.g., tillage versus no-tillage, and how rapidly they change. Several enzymes have shown sensitivity in reflecting early changes in soil quality due to soil management or to pollution long before there are measurable changes in total soil organic C levels (Powlson et al., 1987). Enzyme assays can detect the level of soil degradation (Beck and Beck, 2000; Dick et al., 1996).

Assays on soils polluted with heavy metals have shown that relatively high rates of soil contamination with Zn (300 ppm), Cu (50 ppm), and Cd (3 ppm) reduced enzyme activities involved in C, N, P and S cycling and that soil enzymes have the potential to help us assessing the bioavailability of metals (Kandeler et al., 2000).

Techniques for determining microbial communities and diversities

Phospholipid fatty acid determination (PLFA technique)

The measurement of microbial biomass by fumigation and related techniques is important for its holistic quantification and the determination of microbial C, N, P, and S contents. However, it does not allow determination of the composition of the soil microflora, which can be determined by the profile of phospholipid fatty acids (Zelles et al., 1992; Hamer et al., 2007). The PLFA technique is based on the extraction, fractionation, methylation, and chromatography of the phospholipid fraction of soil lipids (Fig. 1.6).

Bacterial or fungal biomass can be obtained by summing up specific fatty acids (Frostegard and Baath, 1996). On average the conversion factor of total PLFA into microbial biomass C is 5.8 nmol (g soil)$^{-1}$. This factor shows a considerable range from 2.4 to 13.8, with a coefficient of variation of 60%.

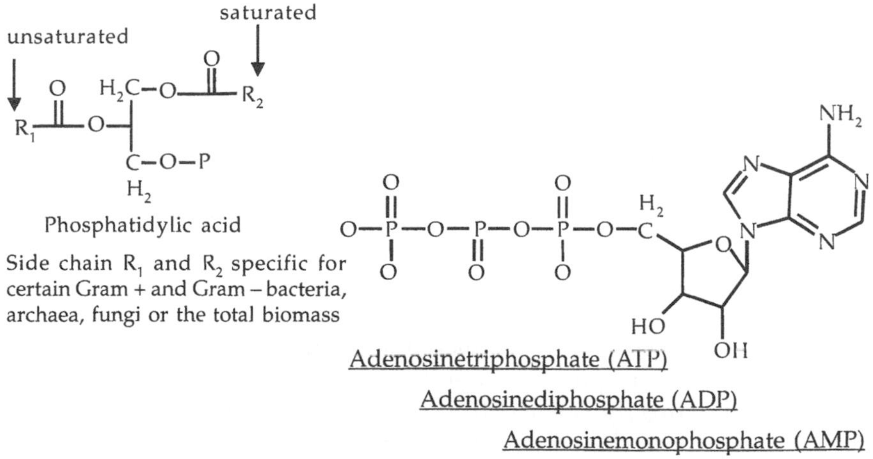

Fig. 1.6: Signature of phospholipids and fatty acids.

BIOLOG —Technique and its problems

The BIOLOG-microtitre plates were originally designed for identification of gram-negative bacteria but can also be used to identify bacteria extracted from soil after appropriate dilution (Winding, 1994).

The BIOLOG method is suitable to measure microbial diversity because it is rapid and simple (Zelles, 1999), but has the drawbacks that it is culture-dependent and that reproducible results can be obtained only if replicates contain identical community profiles and are of similar inoculation density. Furthermore, the contribution of fungi cannot be measured because of their slow growth.

Molecular microbiology methods for characterizing microbial biomass

Traditional methods for the analysis of soil processes involve measuring the distribution of chemical compounds and determining their transformation rates. For example, the changes of nitrogenous compounds in soil over time can be measured via standard techniques, and these data can be used to develop models, e.g., for nitrogen cycling in soil (De Willigen and Neetson,1985). These approaches, however, do not give a complete view into the complexity of the microbial contribution to soil function (Taroncher-Oldenberg et al., 2003).

Molecular approaches based on the small subunits of ribosomal RNA have provided a comprehensive framework for the analysis of microbial communities as shown in Table 1.3 (Kelly, 2003; Amann et al., 1995).

16S-rRNA techniques

The discovery of DNA as the genetic information carrier in all organisms led to the recognition that different species possess unique gene composition. Further advances in the study of bacteria on a molecular base resulted from comparison of the variation in ribosomal RNA (rRNA) by Woese and his group. They compared variations to discern evolutionary relationships among soil microorganisms (Woese, 1987; Woese et al., 1990; Stackebrant, 2001). This can be targeted by extracting DNA (Duarte et al., 1998) or 16S rRNA that can be selectively PCR amplified as shown in Fig. 1.7 (Woese, 1987; Ludwig, 2002).

16S-RNA is composed of two subunits, referred to as small and large tRNA, differentiated by their sedimentation coeficient S. The 16S rRNA has been used mostly for procaryotic soil community analysis (Fig. 1.7). The rRNA can be directly isolated from soil and analyzed as such or after amplification using the polymerase chain reaction (PCR). This method is very useful in soil community analysis because an organism does not need to be cultivated (Fig. 1.8) (Amann et al., 1995).

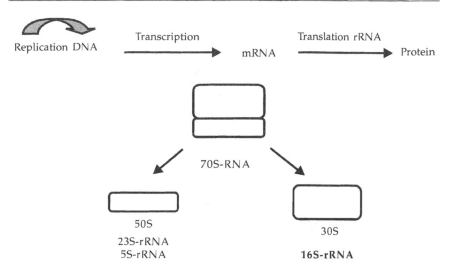

Replication DNA   Transcription ⟶   mRNA   Translation rRNA ⟶   Protein

70S-RNA

50S
23S-rRNA
5S-rRNA

30S
16S-rRNA

Fig. 1.7: 16S-RNA as a species specific marker for bacteria
and it's use as a genetic probe.

Many laboratories are determining 16S rRNA sequences from environmental sources that are added to data banks. Sequence informations and rapidly growing data sets are now available for more 22.000 pro- and eukariontic 16S- or 18S-rRNA (Ludwig, 2002) (Fig. 1.9).

From samples isolated from soil, sediments, or waters, genetic RNA can be extracted. Its amplification by PCR indicated that the rRNA-based tree of life consists of a single domain of eukaryotic organisms and two prokaryotic domains (Archaea and Bacteria).

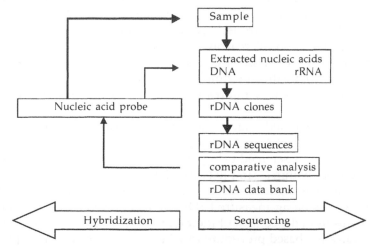

Sample

Extracted nucleic acids
DNA   rRNA

Nucleic acid probe

rDNA clones

rDNA sequences

comparative analysis

rDNA data bank

Hybridization   Sequencing

Fig. 1.8: Steps for rRNA based techniques to evaluate microbial
diversity and communities (Amann, 2000).

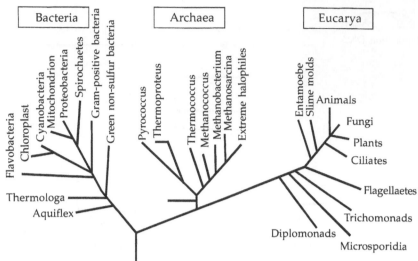

Phylogenetic family tree based on rRNA analyses

Fig. 1.9: Family tree of bacteria based on 16S-rRNA data (Ludwig, 2002).

DNA microarrays

As an alternative to PCR amplification, DNA microarrays offer a much higher probe capacity. They consist of a set of nucleic acids that are spotted and covalently bound onto a solid support such as a glass slide (Smalla et al., 2001). Hundreds to ten thousands of nucleic acids can be spotted within a very small surface area, and this array can be hybridized simultaneously. Recent work has demonstrated that 16S-rRNA extracted from soil can be detected directly without the need for PCR amplification (Smalla et al., 2001; El Fantroussi et al., 2003). Taroncher-Oldenberg et al. (2003) designed a DNA microarray that included oligo-nucleotide probes targeting different variants of the genes involved in the nitrogen cycle. They were extracted from sediment samples from rivers and the extracted DNA was amplified and labeled fluorescently or with a radioisotope (Liebich et al., 2007).

A variety of methods can be used for the analysis of soil microbial communities via 16S-rRNA. A phylogenetic inventory of the prokaryotic component can be assembled using the PCR method to amplify all of the 16S rRNA genes within a sample (Table 1.3).

Therefore, one might be able to determine the actual species involved in the processes being measured (Duarte et al., 1998). Examination of microorganisms in the context of their environment can be made by an in situ hybridization-based procedure in which fluorescently labeled probes diffuse into the cells and are hybridized to the target rRNA and then observed by phase contrast or confocal laser microscopy. General stains such as *acridine orange* (an ethidium bromide that binds to nucleic acids), can be used to detect all organisms. The availability of different colored fluorescent markers makes it possible simultaneously to observe cells hybridized to different probes. Limitations are: background fluorescence of the soil, low cellular ribosome content, and others.

Table 1.3: Several molecular techniques for assessment of microbial communities (Kelly, 2003).

Objective	Target	Techniques
Assignment of phylogenetic diversity	16S-rRNA gene	PCR cloning and sequencing
Detection of phylogenetic groups	16S-rRNA	DNA probes (membrane, in situ or microarray hybridization
Asessment of functional diversity	Functional gene	PCR, cloning, and sequencing
Expression of functional genes	mRNA	DNA probes (membrane, in situ or microarray hybridization)

Functional gene expression in environmental samples is made possible by detecting mRNA as a critical intermediate gene expression and as an indicator that the gene is actively converted to enzymatic protein. Several different techniques have been employed to assess functional gene expression by detecting mRNA. This has been applied to sediments, but not yet to soils (Nogales et al., 2002).

The content of extractable DNA from soil samples can also be used as a measure of microbial biomass with an average conversion factor of 6.0 μg g^{-1} soil. This factor also shows a considerable range from 2.2 to 14.5 with a coefficient of variation of 65%. The ability to extract nucleic acids directly from soil has improved our understanding of the diversity of both cultivated and uncultivated bacteria. We also know that many of the novel soil bacteria cannot be grown in the laboratory, and uncultured bacterial groups are abundant and widespread in soil (Tiedje et al., 1999). Microarrays containing functional gene sequence information are referred to as functional gene arrays, because they are primarily used for analysis of microbial community activities in the environment (Zhou and Thompson, 2002).

REFERENCES

Alef K, Beck TH, Zelles L, Kleiner D, 1988: A comparison of methods to estimate microbial biomass and N-mineralization in agricultural and grassland soils. Soil Biol. Biochem. 20, 561–565.

Amann R, Ludwig W, Schleifer KH, 1995: Phylogenetic identification and in situ detection of individual microbial cells without cultivation. Microbiol. Rev. 59, 143–169.

Amann R, 2000: Microbial aspects of biodiversity. Syst. Apl. Microbiol. 23, 1–8

Anderson JPE, Domsch KH, 1978: A physiological method for the quantitative measurement of microbial biomass in soils. Soil Biol. Biochem. 10, 215–221.

Anderson T, Domsch KH, 1989: Ratios of microbial biomass carbon to total organic carbon in arable soils. Soil Biol. Biochem. 21, 471–479.

Anderson TH, Domsch KH, 1986: Carbon assimilation and microbial activity in soil. Z. Pflanzenern. Bodenk. 149, 457–468.

Anderson TH, Domsch KH, 1990: Application of ecophysiological quotients (qCO_2 and qD) on microbial biomass from soils of different cropping histories. Soil Biol. Biochem. 22, 251–255.

Anderson TH, Weigel HJ, 2003: The current debate about soil biodiversity. Landbauforsch. 53, 223–235.

Anderson TH, 1992: Bedeutung der Mikroorganismen für die Bildung von Aggregaten im Boden. Z. Pflanzenern. Bodenk. 154, 409–416.

Anderson TH , 2008: Assessment of DNA contents of soil fungi. Landbauforsch. 58, 19–28.

Beck T, Beck R, 2000. Kap. 2. 4. 3. 4 "Bodenenzyme" in: Handbuch d. Bodenkunde 8. Ergänzung 2000.

Beck T, 1984a: Der Einfluss unterschiedlicher Bewirtschaftungsmaßnahmen auf die bodenmikrobiologischen Eigenschaften und die Stabilität der organischen Substanz im Boden. Kali Briefe. 17, 331–340.

Beck T, 1984b: Mikrobiologische und biochemische Charakterisierung landwirtschaftlich genutzter Böden. l. Mitt. Die Ermittlung einer bodenmikrobiologischen Kennzahl: Z. Pflanzernähr. Bodenk.147, 456–466.

Beese F, Hartmann A, Beck T, Rackwitz R, Zelles L, 1994: Microbial community structure activity in agricultural soils under different management. Z. Pflanzenern. *Bodenkde.* 157, 187–195.

Brookes PC, Landman A, Pruden G, Jenkinson DS, 1985: Chloroform fumigation and the release of soil nitrogen: A rapid direct extraction method to measure microbial biomass nitrogen in soil. Soil Biol. Biochem. 17, 837–842.

Brookes PC, Newcombe AD, Jenkinson DS, 1987: Adenylate energy charge measurements in soil. Soil Biol. Biochem. 19, 211–217.

Brookes PC, Powlson DS, Jenkinson DS, 1982: Measurement of microbial biomass phosphorus in soil. Soil Biol. Biochem. 14, 319–329.

Burns RG, Dick RP, 2002; Enzymes in the Environment: Activity, Ecology, and Aplications. Marcel Dekker. Inc., New York.

De Willigen P, Neetson JJ, 1985: Comparison of six simulation models for the nitrogen cycle in the soil. Fert. Res. 8: 157–172.

Dennis P, Edwards EA, Liss SN, Fulthorpe R, 2003: Monitoring gene expression in mixed microbial communities by using DNA microarrays. Appl. Environ. Microbiol. 69, 769–778

Dick RP, Kandeler E, 2005: Enzymes in soils. Encycl. Soils in the Environment. Elsevier p. 448–456.

Duarte GF, Rosado AS, Seldin L, Keijzer, Wolters AC, van Elsas JD, 1998: Extraction of ribosomal RNA and genomic DNA from soil for studying the diversity of the indigenous bacterial community. J. Microbiol. Meth. 32, 21–29.

El Fantroussi S, Urakawa S, Bernhard AE, Kelly JJ, Noble PA, Smidt H, Yershov GM, Stahl DA, 2003: Direct profiling of environmental microbial populations by thermal dissociation and analysis of native rRNAs hybridization to oligonucleotideotide microarrays. Appl. Envonm. Microbiol. 69, 2377–2382.

Frostegard A, Peterson SO, Baath E, Nielsen TH, 1997: Dynamics of microbial community associated with manure hot spots as revealed by phospholipid fatty acid analyses. Appl. Environ. Microbiol. 63, 2224–2231.

Gisi U, Schenker R, Schulin R, Stadelmann FX, Sticher H, 1990: Bodenökologie, Thieme, Stuttgart 304 pp.

Haider K, 1996: Biochemie des Bodens. Enke Suttgart 174 pp.

Hamer U, Unger M, Makeschin F, 2007: Impact of air-drying and rewetting on PFLA profiles of soil microbial communities. J. Plant Nutrition Soil Science 170, 259–264.

Kelly JJ, 2003: Molecular techniques for the analysis of soil microbial processes: Functional gene analysis and the utility of DNA microarrays. Soil Sci. 168, 597–605.

Jenkinson DS, Powlson DS, 1976: The effect of biocidal treatments on metabolism in soil. V. A Method for measuring soil biomass. Soil Biol. Biochem. 8, 209–213.

Jenkinson DS, Oades JM, 1979: A method for measuring adenosine triphosphate in soil. Soil Biol. Biochem. 11, 133–199.

Liebich J, Schloter M, Schafer A, Vereecken H, Burauel P, 2007: Degradation and humification Microbial community changes during humification of ^{14}C-labelled maize straw in heat-treated and native Orthic Luvisol: Europ. J. Soil Sci. 58, 141–151.

Ludwig W, 2002: Ribosomale Ribonukleinsäuren als Grundlage für die molekulare Identifizierung der Mikroorganismen. Rundgespr. Komm. Ökol. Bayer. Akad. Wissensch., Bedeut. d. Mikroorg. für die Umwelt 23, 17–30.

Martens R, 1995: Current methods for measuring microbial biomass C in soil: Potential and limitations. Biol. Fert. Soils 19, 87–99.

Martin JP, 1946. Microorganisms and soil aggregation. II. Influence of bacterial polysaccharides on soil structure. Soil Sci. 59: 163–166.

Macrae A, 2000: The use of 16S rDNA methods in soil microbial ecology. Braz. J. Microbiol. 31, 77–82.

McLaren AD, 1975: Soil as a system of humus and clay immobilized enzymes. Chem. Scr. 8, 97–99.

Millar WN, Casida LE, 1970: Evidence for muramic acid in soil. Can. J. Microbiol. 16, 299–304.

Nannipieri P, Ascher J, Landi L, Pietramellara G, Renella G, 2003: Microbial diversity and soil functions. Europ. J. Soil Sci. 54, 655–670.

Nogales B, Timmis BKN, Nedwell DB, Osborn AM, 2002: Detection and diversity of expressed denitrification genes in estuarine sediments after reverse transcription-PCR amplification from mRNA. Appl Environ. Microbiol. 68, 5017–5025.

Paul EA and Clark FE, 1989: Soil Microbiology and Biochemistry. First Edition. Academic Press, San Diego, Cal., 273 pp.

Paul EA and Clark FE, 1996: Soil Microbiology and Biochemistry. Second Edition. Acad. Press, San Diego. Cal., 340 pp.

Powlson DS, Brookes PC, Christensen BT, 1987: Measurement of microbial biomass provides an early indication of changes in total soil organic matter due to straw incorporation. Soil Biol. Biochem. 19, 159–164.

Schadt CW, Liebich J, Chong S Cn Gentry TJ, Zhili H, PanH, Zhou J, 2005: design and use of functional gene microarrays (FGAs) for the characterization of microbial commuities. Meth Microbiol. 34, 33–368.

Schinner F, Öhlinger R, Kandeler E, 1996: Methods in Soil Biology, Springer, Berlin 426 pp.

Sexstone AJ, Reusbech NP, Parkin TN, Tiedje JM, 1985: Direct measurement of oxygen profiles and denitrification rates in soil aggregates. Soil Sci. Soc. Am. J. 49, 445–451.

Smalla J, Call DR Brockman FJ, Straub TM., Chandler P, 2001: Direct detection of 16S rRNA in soil extracts by using oligonuleotide microarrays. Appl. Environ. Microbiol. 67, 4708–4716.

Sörensen LH, 1974: Rate of decomposition of organic matter in soil as influenced by repeated air drying-rewetting and repeated additions of organic material. Soil Biol. Biochem. 6, 287–292.

Stackebrand E, 2001: Fortschritte in der Systematik der Bakterien. Naturwissensch. Rundschau 54, 345–354.

Stott DE, Kassim G, Jarrell WM, Martin JP, Haider K, 1983: Stabilization and incorporation into biomass of specific plant carbon during biodegradation in soil. Plant Soil 70, 15–26.

Stout, JD, 1980: The role of protozoa in nutrient cycling and nutrient turnover. Adv. Microb. Ecol. 4, 1–50.

Tabatabai M, Fung M, 1992: Extraction of enzymes from soil. In: Soil Biochemistry, G. Stotzky and J.M. Bollag (eds.) Vol 7, 197–227 pp. Dekker, New York.

Taroncher-Oldenberg G, Griner E, Francis M, Ward BB, 2003: Oligonucleotide microarray for the study of functional gene diversity in the nitrogen cycle in the environment. Appl. Environ. Microbiol. 69, 1159–1171.

Tiedje JM, Assuming-Brempong S, Nusslein K, Marsh TL, Flynn SJ, 1999: Opening the black box of soil microbial diversity. Appl. Soil Ecol. 13,109–122.

Torsvik V, Goksyor J, Daae R, Sörheim R, Michalsen J, Salte K,1994: Use of DNA analysis to determine the diversity of microbial communities. In: Beyond the biomass. Compositional and functional analysis of soil microbial communities. K. Ritz et al. (eds.) J. Wiley and Sons, Chichester pp. 39–48.

Tunlid A, White DC, 1992: Biochemical analysis of biomass, community structure, nutritional activity of microbial communities in soil. In: Soil Biochemistry Vol. 7., G. Stotzky and J.M. Bollag (eds.), M. Dekker, New York, p. 229–262.

Winding A, 1994: Fingerprinting bacterial soil communities using Biolog microtitre plates, In: Beyond biomass, eds. K. Ritz, J. Dighton and K.E. Giller, pp. 85–100, Wiley-Sayce Publication, Chichester.

Woese CR, 1987: Bacterial evolution, Microbiol. Reviews 51, 221–271.

Woese CR, Kandler O, Wheelis ML, 1990: Towards a natural system of organims: proposal for the domains Archaea, Bacteria and Eucaria. Proc. Natl. Acad, Sci. USA 87, 4576–4579.

Zelles L, 1999: Fatty acid pattern of phopholipids and lipopolysaccharides in the characterisation of microbial communities in soil: a review. Biol. Fert. Soils 29, 111–129.

Zelles L, Bai QY, Beck T, Beese F, 1992: Signature fatty acids in phospholipids and lipopolysaccharides as indicators of microbial biomass and community structures in agricultural soils. Soil Biol. Biochem. 24, 317–323.

Zhou J, Thompson DK, 2002: Challenges in applying microarrays to environmental studies, Curr. Opin. Biotechnol. 13, 204–207.

Aerobic and Anaerobic Degradation of Monomer and Polymer Plant Constituents by Soil Microorganisms

Soil microorganisms finally degrade any naturally occurring compounds from plant residues into mineral constituents. This guarantees the function of the natural cycles and stops the accumulation of organic materials. Microorganisms have developed multivariant strategies for degradation of any compound. Even brown coal or mineral oil and many xenobiotics can be degraded. Their accumulation in deposits results from the absence of suitable oxidizing agents, especially of oxygen.

Degradation and transformation of organic compounds and particularly of natural polymers too large for direct uptake by organisms for assimilation can be divided into three phases (Fig. 2.1).

High molecular weight substances such as proteins, polysaccharides, and similar compounds are first cleaved into smaller products that can be absorbed by the cells and further metabolized (Fig. 2.2). This occurs by different modes of glycolysis, oxidation, and transformation of the resulting C_2 bodies such as acetate by the tricarboxylic acid cycle (TCA cycle) to CO_2 and hydrogen. This cycle also produces energy and building blocks from compounds that, because of their molecular size, cannot be directly used by higher organisms.

Cellulose and lignin are the main shape- and structure-giving compounds of plants. They both are the most abundant biopolymers on earth.

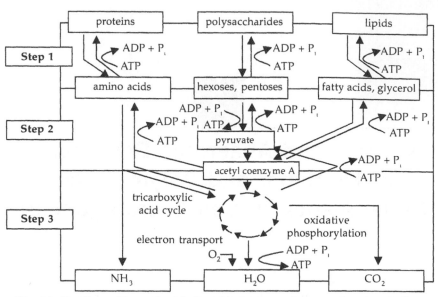

Fig. 2.1: The three phases of catabolism (dark arrows downward) and anabolism (light arrows upward) (according to Lehninger, 1996).

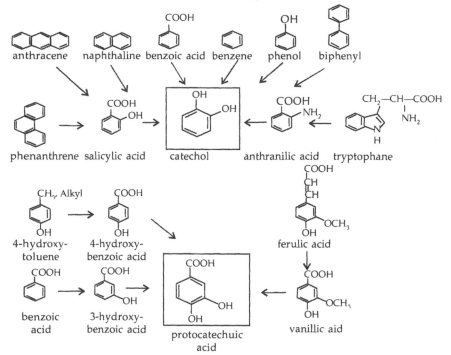

Fig. 2.2: Degradation of aromatic structures with catechol and protocatechoic acid as central metabolites.

Transformation of carbohydrates and polysaccharides

Cellulose globally amounts to about 40×10^9 tons annually formed and ultimately degraded by the enzymatic activity of microbes. Cellulose forms bundles or microfibrils in which the cellulose molecules are oriented and are held together by hydrogen bonds. They have crystalline and more disordered, amorphous regions. Cellulose fibrils in the plant cell walls are embedded in a matrix of hemicelluloses and lignin (Fig. 2.3, 2.4, 2.5, 2.6, 2.7).

A model that accounts for the presence of polyoses was proposed by Fengel and Wegener (1984) explaining the intimate association of cellulose and polyoses and of lignin in the cell wall.

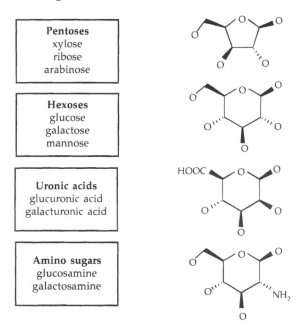

Fig. 2.3: Structural units of polyoses (hemicelluloses).

The soil surface is the most important aerobic environment where dead plant material (wood, leaves, straw, etc.) accumulates and degrades. The most recalcitrant substrate is wood, which is highly lignified. White rot fungi from the basidiomycete family can degrade both lignin and cellulose and plays a major role in wood decay (Kirk, 1984).

Other aerobic fungi are deficient in lignase activity, but are efficient cellulose degraders with an extensively studied cellulase system. A number of actinomycetes, including nocardiae and rhodococci, are active but

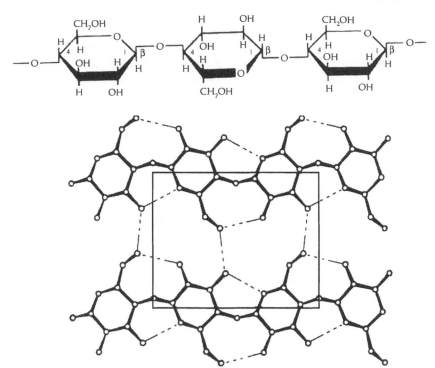

Fig. 2.4: Polymeric cellulose structure stabilized by H-bonding.

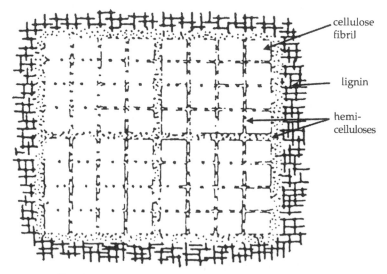

Fig. 2.5: Development of a lignified cell wall (Fengel and Wegener, 1984).

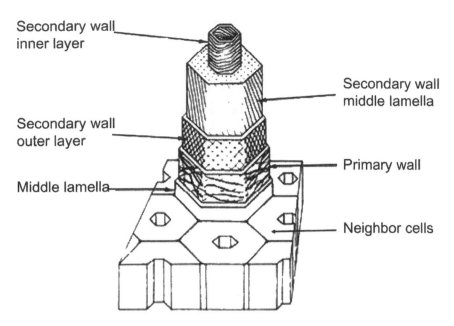

Fig. 2.6: Structure of the lignified plant cell wall (Fengel and Wegener, 1984).

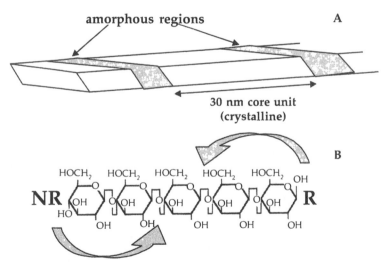

A: Cellulose microfibril: crystalline segments (white) are interrupted by amorphous segments (black)

B: Exoglucanases degrade fragments of cellulose from the reducing (R) and non-reducing ends (NR)

Fig. 2.7: Structural features of cellulose fibrils in the cell wall
(according to Schwarz, 2003).

slower in lignin degradation than white rot fungi, and some cellulolytic and thermophilic actinomycetes such as *Thermomonospora* and *Microbispora* can attack lignocellulose.

Biochemistry of cellulases

The conversion of cellulosic materials into glucose is of considerable complexity and the matrix of hemicellulose and lignin severely restricts the access of cellulolytic enzymes (Fig. 2.6). Their enzymes can be found as individual proteins or are physically associated to form high molecular weight complexes, the cellulosomes (see 2.1.2).

The main features of the cellulase system are indicated in Fig. 2.8 and 2.9.

Endoglucanases act on the non crystalline regions of cellulose and cleave randomly the long glycosidic chains; cellobiohydrolases cleave dimers from the nonreducing end, which are then hydrolzed to yield glucose (black hexagons). Many of the features of cellulose degradation also have been demonstrated for other aerobic fungi and several cellulolytic bacteria (e.g. *Cellulomonas* sp., *Streptomyces* sp., *Clostridium* sp. and others) containing endoglucanases and β-glucosidases. The synergy between endo- and exoglucanases and β-glucosidases in the degradation of native cellulose (Fig. 2.6) is also central to the understanding of cellulase

Basisdiomycetes
White and brown rod fungi

Ascomycetes and Fungi imperfecti
Aspergillus, Trichoderma, Penicillium, Fusarium, Verticillium

Bacteria (aerobic)
Cellulomonas, Bacillus, Cellovibrio, Streptomyces, Thermoactionomyces, Pseudomonas, Cytophaga, Sporocytophaga

Bacteria (anaerobic)
Clostridium, Bacteroides, Micromonospora

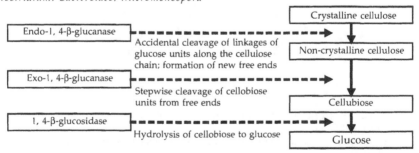

Fig. 2.8: Examples of cellulose degrading organisms: fungi, aerobic and anaerobic bacteria (Haider, 1996, according to Wagner and Sistig, 1979).

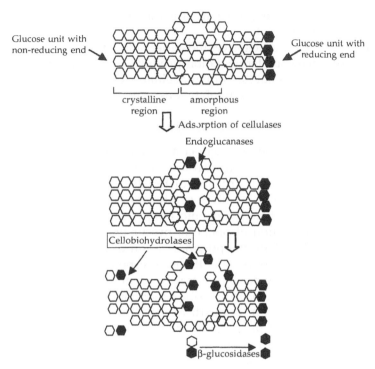

Fig. 2.9: Schematic representation of *Trichoderma reesei* cellulases (Beguin and Aubert, 1992).

systems of several anaerobic microorganisms having cellulase systems in which the individual components are associated in high-molecular weight complexes termed **cellulosomes**.

Current ideas about the structure and mode of action of cellulosomes are based on electron micrographs showing clusters of subunits arranged in parallel to their major axis. At least 14 different components ranging from 40 to 250 kDa can be identified by gel electrophoresis. The cellulosome binds strongly to cellulose by a subunit; another subunit aligned at closely spaced intervals cleaves a single cellulose molecule at multiple sites and the hydrolysis products are further processed and metabolized. Features of the complicated *Cellulomonas thermocellum* cellulolytic system in rumen are displayed by a number of cellulolytic bacteria, including *Clostridium cellobioparum, Rhuminicoccus albus*, and other *Cellulomonas* spp.

Cellulose degradation by organisms in natural habitats is done by mixed populations including cellulolytic and noncellulolytic microorganisms that interact synergistically with formulation of CO_2, CH_4, and H_2O in anaerobic environments. In the rumen system anaerobic fungi are associated with cellulolytic bacteria, and fermentation products formed are acetate, lactate, formate, succinate, and others, which are partly used as food or are transformed into CH_4.

In most anaerobic systems cellulose degradation seems to be mainly due to bacteria, the main fermentation products are lactate, ethanol, CO_2, and H_2. Due to the insolubility of lignocelluloses their compact structure restricts the accessibility of cellulolytic enzymes.

Degradation and transformation of lignin

Lignin like cellulose is another major component of plant material and the most abundant form of aromatic carbon in the biosphere. It provides a barrier against microbial destruction of the degradable polysaccharides in the plant cell walls. Chemically it is a heterogenous, optically inactive polymer consisting of phenylpropanoid units (Boudet, 2000), that arise by enzymatic polymerization of phenolic precursors by peroxidases, and phenolases, which initiate the polymerization of *coniferyl, sinapyl,* and *p-coumaryl* alcohols via coupling of their corresponding phenoxy radicals to a polymer linked together by several covalent bonds (aryl-ether, aryl-aryl, carbon-carbon bonds), as described by Boudet (2000) and shown in Fig. 2.10 and 2.11. Lignin deposition in plant cell walls allows the development of upright plants in the terrestrial environment. It is after cellulose the second most abundant plant biopolymer and accounts for about 30% of plant biomass.

The process of enzymatic lignin formation was named by Freudenberg as "dehydrogenative phenol polymerization" (DHP-lignin). Infrared-spectroscopy revealed that the identical nature of both "artificial" lignin and lignin in the cell wall showed differences in carbonyl contents, in ethylenic double bonds, and in contents of ether and hydroxyl groups. ^{13}C NMR spectral analysis and thioacidolysis also showed that artificial and natural lignin are almost identical in terms of bond types and polymer size (Lüdemann and Nimz, 1974).

Due to its multiple types of bonds and their heterogeneity, lignin cannot be degraded by hydrolysis as in case of cellulose, proteins and other natural polymers. Lignin degradation is brought about by a group of white rot basidiomycete fungi that have developed an efficient enzyme system to selectively remove lignin, while leaving white cellulose fibers. These fungi are therefore also called white rot fungi (Table 2.1); typical white rot fungi are *Trametes versicolor, Phanerochaete chrysosporium, Pleurotus ostreatus, Nematoloma frowardii.*

Lignin degradation by white rot fungi is brought about by a synergistic cooperation of phenol-oxidizing enzymes, manganese peroxidase, lignin peroxidase and laccase which are necessary to attack lignin (see Table 2.1 and Fig. 2.12) (Hatakka, 1994). Both peroxidases are ferric iron containing heme proteins and laccase belongs to the copper-containing blue oxidases.

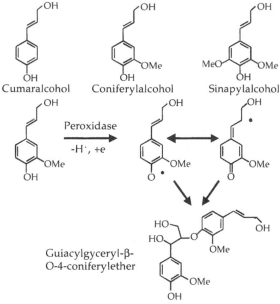

Cumaralcohol   Coniferylalcohol   Sinapylalcohol

Peroxidase

-H·, +e

Guiacylgyceryl-β-
O-4-coniferylether

Fig. 2.10: Lignin formation (according to Boudet, 2000).

Table 2.1: Several ligninolytic microorganisms from different microbial groups (Haider, 1996). Methods of investigation: A = Degradation of lignocelluloses (wood or straw); B = mineralization of lignin or of DHP-lignin; C = splitting of methoxyl groups or side chains; D = degradation of phenolic lignin monomers.

Microorganisms (classification)	Methods of investigation	Degree of degradation
Basidiomycetes		
White rot fungi		
Phanerochaete chrysosporium	A, B, D	Intense
Pleurotus ostreatus	A, B, D	Intense to medium
Coriolus versicolor	A, B, D	Intense
Phlebia radiate	A, B	Intense to medium
Dichomitus sp.	A	Intense
Brown rot fungi		
Lentinus lepideus	A, B, C, D	Medium to weak
Poria placenta	B, C	Weak
Fungi imperfecti, Deuteromycetales		
Soft rot fungi		
Chaetomium globosum	A, B, C, D	Medium
Preussia fleishhakii	B, C, D	Medium
Paecilomyces sp.	A, B, C	Medium
Graphium sp.	A, B, C	Medium
Bacteria and actinomycetes		
Streptomyces viridosporus	A, B, C	Medium
Nocardia autotrophica	B, C, D	Medium to weak
Bacillus sp.	B, C	Weak
Xanthomonas sp.	B, C	Medium to weak

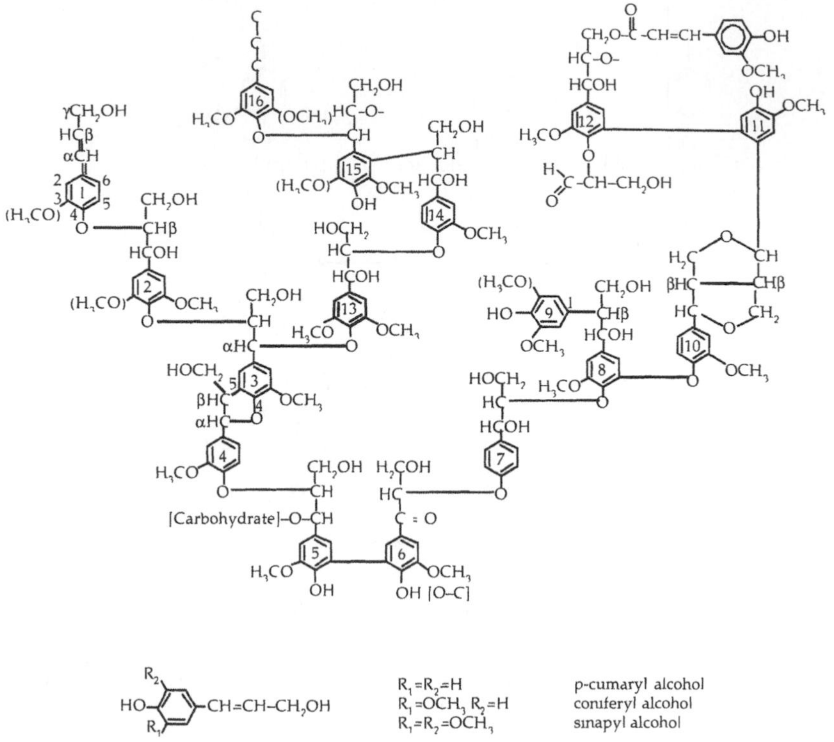

Fig. 2.11: Structure of coniferous lignin (according to Freudenberg and Neish, 1968).

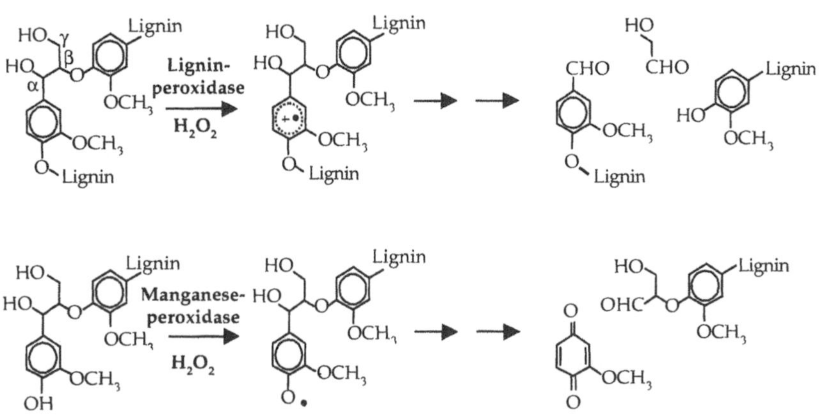

Fig. 2.12: The microbial degradation of lignin (Kirk and Farrel, 1987).

Lignin degraders have peroxidases and laccase that catalyze biological oxidation of lignin by free oxygen radicals. Therefore, lignin cannot be degraded under anaerobic conditions and accumulates in peat and brown coal (Kirk and Farrel, 1987). Even under aerobic conditions where lignin is mineralized completely, its carbon and energy can not be used by the ligninolytic microorganisms for metabolic reactions or growth (Haider, 1992).

Lignin degradation and humus formation

Lignin has long been suspected to be an important source of stable carbon in soils and in humus formation. Lignin inputs from plant residues are large and because it is the second most important plant constituent after cellulose (Crawford, 1981). Due to its polyphenolic structure lignin is considered more recalcitrant to biochemical degradation than most other families of plant molecules. A two-reservoir model with lignin in undecomposed plant residues and another bigger pool with lignin residues protected in the soil matrix was observed and proposed by Stott et al. (1983) and later on by Rasse et al. (2006).

Lignin solubilization also occurs in termite guts. Lignin solubilising *Streptomyces* spp. have been shown to oxidatively depolymerize lignin and to produce a water-soluble acid-precititable lignin degradation product that is similar to soil humus.

Degradability of plant cell walls by ruminants

The degradation of plant cell walls by ruminants is of major importance for animal and human nutrition. Increasing the efficiency with which the rumen microbiota degrade the fiber has been the subject of extensive research (e.g., Krause et al., 2003).

Enhancement of digestibility of plant fibers by rumen microorganisms

Lignification reduces the degradability of plant cell walls in the rumen of ruminants, and there are a number of technologies to increase the digestibility based on mechanical and chemical treatments or by the introduction of transgenic plants. Degradation of plant cell walls and cellulose-rich material occurs in aerobic as well as in anaerobic habitats, the latter of which include the rumen and intestinal tract of animals, sewage sludge digestors, composts, fresh- and seawater muds, and sediments.

Lignocelluloses are important in ruminant diets. Improvement in the ability of the rumen to degrade plant cell walls are aimed by reducing lignin contents. This can for example be achieved by a previous incubation of straw with ligninolytic fungi. Agosin et al. (1985) showed that *in vitro* digestibility of wheat straw increased from 38 to 68% when treated previously with strains of *Cyathus stercoreus* or other ligninolytic fungi (see Table 2.1). However, prolonged incubation with these organisms resulted in carbohydrate degradation and thereby reduced digestibility.

A number of other technologies increase the digestibility of fibers in the rumen based on mechanical and chemical treatments of plant material before it is consumed. They are mostly designed to change the relationship between lignin and carbohydrates within the plant cell wall by grinding or hydrolysis (Sun and Cheng, 2002).

Recently molecular strategies for manipulating the enzymes of lignin biosynthesis contributed to our understanding of the process of lignin biosynthesis and the effects of altered lignin content and composition on degradability in soil and in the rumen (Whetten et al., 1998). Improvement in digestibility was observed by molecular strategies in several forage plants with reduced and altered lignin contents. The enzymes that catalyze lignin formation in plants are encoded by multigene families that generate a number of related reactions; consequently the effects of suppressing single enzymic reactions are not predictable. Therefore, experiments into the digestibility showed conflicting results. Transgenic tobacco plants with reduced expression of cinnamaldehyde dehydrogenase produced lignin with a decreased syringyl/guaiacyl ratio and increased degradability even though there was no change in total lignin content. In other cases experiments with transgenic plants showed that increased digestibility was associated with reduced lignin content, not with changes in composition.

Degradation of organopollutants by ligninolytic organisms or enzymes

The unique ligninolytic enzyme system of basidiomycete fungi, which is based on a highly reactive free radical depolymerization mechanism, is ideal for the biodegradation of polycyclic organopollutants in the environment (Fernando et al., 1990; Bumpus, 1993), e.g. benzo(a)pyrene and other polyaromatic hydrocarbons by *Bjerkandera adusta* and *Nematoloma frowardii* (Sack et al., 1997; Sack and Fritsche, 1997; Scheibner et al., 1997). These fungi also degrade humic compounds with bound residues from pesticides (Haider and Martin, 1998) (Table 2.2).

Table 2.2: Biotransformation catalyzed by white rot fungi and by lignin peroxidase enzymes (Glaser and Lamar, 1995).

Reaction	Structural change
Aromatic ring oxidation	$ArH \rightarrow ArOH$
Aryl ether oxidation	$ArOR \rightarrow AROH$
Diol cleavage	$ArCH(OH)CH_2OH \rightarrow ArCHO + HCHO$
Benzylic methylene hydroxylation	$ArCH_2R \rightarrow ArCR(OH)R$
Styryl olefin hydroxylation	$ArCR=CCR_2 \rightarrow ArCR(OH)CR_2OH$
Benzyl alcohol oxidation	$ArCH_2OH \rightarrow ArCHO$
Oxidative coupling	$ArH + Ar'H \rightarrow Ar-Ar'$

Field application of *Phanerochaete* spp. and other lignin-degrading *Phanerochaete* strains to eliminate pentachlorophenol (PCP) contamination from soil has been encouraging; PCP concentrations of 600 mg kg^{-1} have been reduced by 55 to 84% over a 56-day period in topsoil tests (Lamar et al., 1990). Several recalcitrant organopollutants such as DDT, TNT, and others, can be dehalogenated or degraded by ligninolytic organisms.

REFERENCES

Adler E,1977: Lignin chemistry. Past, present, future. Wood Sci. Technol. 11: 169–218.

Agosin E, Monties B, Odier E, 1985: Structural changes in wheat straw components during decay by lignin-degrading white-rot fungi in relation to improvement of digestibility for ruminants, J. Sci. Food Agric. 36, 925–935.

Beguin P, Aubert, JP, 1992: Cellulases. Encycl. Microbiol. Vol 1: 467–477. Acad. Press, New York.

Boudet AM, 2000: Lignins and lignification: Selected issues. Ann. Rev. Plant Physiol Biochem. 38, 81–96.

Bumpus JA, 1993: White rot fungi and their potential use in bioremediation processes. Soil Biochemistry, Vol 8 (JM Bollag and Stotzky (eds.), pp. 65–100. M. Dekker, NewYork.

Chen Y, Tarchitzky J, Markovitch O, 2006: Organic matter transformations during the composting of biosolids. Proceedings of the 13th Meeting of the International Humic Substances Society, Karlsruhe.

Crawford RL, 1981: Lignin Biodegradation and Transformation. Wiley-Interscience, New York, 154 pp.

Demain AL, Wu JHD, 1989: The cellulase complex of Clostridium thermocellum. In T.K. Ghose (ed.): Bioprocess engineering, Ellis Horwood, Chichester, p 68–86.

Dignac MF, Bahri H, Rumpel C, Rasse DP, Bardoux G, Balesdent J, 2005: Carbon–13 natural abundance as a tool to study the dynamics of lignin monomers in soil. Geoderma 128:1–17.

Fengel D, Wegener G, 1984: Wood: Chemistry, Ultrastructure, Reactions. De Gruyter, Berlin, 613 pp.

Fernanando T, Bumpus JA, Aust SD, 1990: Degradation of TNT by Phanerochaete chrysosporium. Appl. Env. Microbiol. 56, 1666–1671.

Freudenberg K, Neish AC, 1968: Constitution and biosynthesis of lignin. Springer, Berlin, Heidelberg, New York, 129 pp.

Glaser JA, Lamar RT, 1995: Lignin-degrading fungi as degraders of pentachlorophenol and creosote in soil. In: Bioremediation, Science and Applicationns. H.D. Skipper and R.F. Turco eds., Soil Science America Special Publication 43, Madison WI, pp. 117–134.

Haider K, 1992: Problems related to the humification processes in soils of temperate climate. In: Soil Biochemistry, Vol 7, G. Sotzky, J.-M. Bollag (eds.), Marcel Dekker, Incorp. pp. 55–94.

Haider K, 1996: Biochemie des Bodens. Enke, Stuttgart, 174 pp.

Haider K, Martin JP, 1988: Mineralization of C-14 labelled humic acids and of humic acid bound C-14 xenobiotics by *Phanerochaete chrysosporium.* Soil Biol. Biochem. 20, 425–429.

Haider K, Martin JP, 1988: Mineralization of ^{14}C-labeled humic acids and humic-acid bound ^{14}C-xenobiotics by *Phanerochaete chrysosporium.* Soil Biol. Biochem. 20: 425–429.

Hatakka A, 1994: Lignin modifying enzymes from selected white-rot fungi: production and role in lignin degradation. FEMS Microbiol. Rev. 13, 125–135.

Kirk TK, 1984: Degradation of Lignin: In: Microbial degradation of organic compounds. D.T. Gibson (ed) Marcell Dekker, New York, pp. 399–438.

Kirk TK, Farrel TI, 1987: Enzymatic combustion—the microbial degradation of lignin. Annu. Rev. Microbiol. 41, 465–505.

Krause DO, Denman SE, Mackie RI, Morrison M, Rae AL, Aewood GT, Sweeney CS, 2003: Opportunities to improve fiber degradation in the rumen: microbiology, ecology, and genomics. FEMS Microbiology Rev. 27–663–693.

Lamar RT, Davis MW, Dietrich DM, Glaser JA, 1994: Treatment of a Pentachlorophenol Contaminated and Creosote Contaminated soil using the Lignin-degrading Fungus Phanerochaete-sordida—A Field Demonstartion. Soil. Biol. Biochem. 26, 1603–1611.

Lamar RT, Glaser JA, Kirk TK, 1990: Fate of pentachlorophenol (PCP) in sterile soils inoculated with white rot basidiomycete Phanerochaete chrysosporium. Soil Biol. Biochem. 22: 433–440.

Lehninger AL, 1996: Principles of Biochemistry. Spectrum, Heidelberg, 1074 pp.

Lüdemann HD, Nimz H, 1974: 13C-Kernresonanzspektren von Ligninen. 2. Buchen- und Fichten-Björkman Lignin. Makromol. Chem. 175:2409–2422.

Rasse DF, Dignac MF, Bahri H, Rumpel C, Mariotti A, Chenu C, 2006: Lignin turnover in an agricultural field: from plant residues to soil-protected fractions. Europ. J. Soil Sci. 57:530–538.

Sack U, Hofrichter M, Fritsche W, 1997: Degradation of polycyclic aromatic hydrocarbons by manganese peroxidase of Nematoloma frowardii. FEMS Lett. 152, 227–234.

Scheibner K, Hofrichter M, Herre A, Michels J, Fritsche W, 1997: Screening for fungi intensively mineralizing 2,4,6-trinitrotoluene. Appl. Environ. Microbiol Biotechnol. 47, 452–457.

Schwarz WH, 2003: Das Celllulosom—eine Nano-Maschine zum Abbau der Cellulose. Naturwiss. Rundschau 5: 121–128.

Sewalt VJH, Beauchemin KA, Rode LM, Acharya S, Baron VS, 1997: Lignin impact on fiber degradation. 4. Enzymatic saccharification and in vitro digestibility of alfalfa and grasses following selective solvent delignification. Biores. Technol. 61, 199–206.

Stott DE, Kassim G, Jarrell WM, Martin JP, Haider K, 1983: Stabilization and incorporation into biomass of specific plant carbon during biodegradation in soil. Plant Soil 70: 15–26.

Sun Y, Cheng J, 2002: Hydrolysis of lignocellulosic materials for ethanol production: a review. Bioresour. Technol. 83, 1–11.

Wagner F, Sistig P, 1979: Verwertung von Cellulose durch Mikroorganismen. Forum Mikrobiologie 2: 74–80.

Whetten RW, MacKay JJ, Sederoff RR, 1998: Recent advances in understanding of lignin biosynthesis. Ann Rev. Plant Physiol. 49, 585–609.

3

Humus and Humification

Humus properties and development

The content of humus and its quality are of great significance for soil fertility and plant production and influence numerous chemical and physical soil properties listed in Table 3.1.

Table 3.1: Fortunate properties of soil humus (see also Hayes et al., 1989).

— Sorption and/or complexation of plant nutrients and trace elements; their slow and timely release
— Maintenance of good soil structure; promotion of aggregate formation; improvement of porosity
— Improvement of water and air movement, and water holding capacity, and water retention
— Enhancement of the filter and buffer capacity
— Immobilization and decontamination of organic and inorganic toxic compounds
— Promotion of plant growth by specific compounds (phytohormones)

Soil organic matter (SOM) is derived from dead plant residues consisting of lignocelluloses and develops from these polymers by microbial activity as a natural product in the inorganic/organic soil environment. The primary sources of SOM are dead plant materials, such as straw, twigs, roots and other plant litter materials. Soils are major players in the global carbon cycle. On a global scale (Table 3.2) they store the equivalent of about 300 times the amount of carbon annually released through the burning of fossil fuels, and they amount to about 2.2×10^{18} g C distributed on a land area of 150×10^{12} m^2 (Batjes, 1996).

The global biomass production amounts to approximately 500 to 700 Gt C (Gt = 10^9 t) of higher plants consisting of lignocelluloses with an average composition of 15 to 60% cellulose, 10 to 30% hemicelluloses (polyoses), 5 to 30% lignin, and 2 to 15% proteins together with minor amounts of phenols, amino acids, peptides, and sugars, as well as numerous products of the secondary metabolism.

The greater part of this starting material is immediately used by microorganisms as nutrient and energy sources and converted to CO_2 evolving from the soil surface. This annual CO_2 surface flux is estimated approximately at 75×10^{15} g C/yr^{-1} by rhizomicrobial respiration (Schlesinger, 1997; Hanson et al., 2000).

Hedges (1988) considered humification to be a slow oxidation, condensation and decomposition process of plant biopolymers in which the parent material is not completely destroyed by microbial degradation. From NMR studies, Simpson et al. (2003) similarly suggested that SOM formation is a consequence of the transformation of biopolymers by microbial action and not a result of random condensation reactions. Accumulation of SOM is controlled by the composition and amounts of plant residues, by climatic conditions, and by soil texture.

Table 3.2: Distribution of soil organic carbon and plant biomass in various global ecosystems (Schlesinger,1991).

Type of ecosystem	Soil org. C kg C m^{-2}	Area 10^8 ha	Total soil org. C 10^6 g C ha^{-1}	Plant biomass 10^6 g C ha^{-1}	Net primary production 10^6 g C ha^{-1} yr^{-1}
Tropical forest	10.4	24.5	255	19	19
Temperate forest	11.8	12	142	12	12
Boreal forest	14.9	12	179	No data	No data
Woodland/shrubland	6.9	8.5	59	No data	No data
Tropical savannah	3.7	15	56	No data	No data
Temperate grassland	19.2	9	173	0.7	6
Tundra/alpine	21.6	8	173	No data	No data
Desert shrub	5.6	18	101	0.01	1
Cultivated	12.7	14	178	7	6
Swamp/marsh	68.6	2	137	No data	30
Total world	No data	147	1570	560	No data

Plant residues introduced into soil are the main source of humus. During soil incubation in a temperate climate, about 70% of the total residue carbon is released as CO_2. The remainder decomposes more and more slowly with time and becomes steadily incorporated into soil humus. Although complex processes are involved during decomposition of the various ingredients, the overall degradation rates follow reasonably (Jenkinson and Rayner, 1977) well first-order kinetics under both laboratory and field conditions.

$$A_t = A_0 \cdot e^{-kt}$$

with A_0 being the amount of organic matter at the beginning and A_t being the amount at a distinct time and k being the rate constant (Table 3.3).

Table 3.3: Degradation of various organic compounds and plant residues upon laboratory incubation in a Greenfield Sandy Loam Topsoil[1] (Martin and Haider, 1986).

Substrate	Degradation after weeks (% of applied carbon)[2]					
	1	4	12	20	28	40
Glucose	73	82	89	90	90	91
Starch	48	69	81	84	86	88
Cellulose	27	52	77	79	84	86
Green manure (corn 86 days)	27	45	69	72	82	83
Wheat straw (ripe)	20	33	59	61	64	66
Cow dung (dry)	18	33	43	48	50	51
Saw dust (plum wood)	12	25	33	40	45	47
Saw dust (pine wood)	2	5	14	29	34	35
Peat (dry)	<1	3	8	14	17	18
Humic acids, Chernozem (dry)	<1	<1	1	1	1	2

1. Dry powdered (1 mm) material was thoroughly mixed at 1000 ppm with dry soil and incubated at −33 kPa water potential at 22 °C in the laboratory under continuous aeration.
2. Per cent of the added carbon evolved as CO_2.

From the results of a now 100-years-running field experiment with regular annual straw fertilization corresponding to 1,000 kg C ha^{-1} yr^{-1} Jenkinson and Rayner (1977) concluded that decomposition and humification can best be described by a bipartite logarithmic e-function (first order kinetics two compartment)

$$A_t = A_{01} \cdot e^{-k_1 t} + A_{02} \cdot e^{-k_2 t}$$

where the first part describes the decomposition of the more easily decomposing polysaccharide portion and the second part the more recalcitrant lignin portion. At equilibrium of addition and decomposition in the upper 23 cm of soil, 24 t of newly formed "stabilized" humus mostly sorbed to heavy metal oxides was formed.

Whereas previous studies (Sauerbeck and Gonzalez, 1977; Jenkinson, 1977) used radiolabelled (^{14}C-labelled) plant materials for their investigations, more recent studies apply the natural ^{13}C labeling of SOM generated from chronosequences of 9 years maize as a C4 crop replacing the previous wheat C3 crop. Here the combined applications of CuO oxidation and gas chromatography coupled via a combustion interface to an isotope ratio mass spectrometer (GC/C-IRMS =Gas-Chromatography/Carbon-Isotopic Ratio Mass Spectrometry) allowed Dignac et al. (2005) to follow variations in the isotopic signature of lignin derived monomers.

Soil fabric and its impact on SOM sequestration

Anderson et al. (1981), using ultrasonic dispersion and size fractionation techniques, indicated, that 96% of the organic C and 93% of the organic N was associated with mineral soil particles ranging from sand to the fine clay fraction (Table 3.4).

Table 3.4: Distribution of organic carbon, nitrogen, carbohydrates, and C/N ratios among particle size fractions in a black Chernozem soil (Anderson et al., 1981).

Fraction	Whole soil 2 mm	Sand 0.05–2 mm	Coarse silt 5–50 μm	Fine silt 2–5 μm	Coarse clay 0.2–2 μm	Fine clay <0.2 μm	Recovery %
Organic C%	3.30	3.9	25.0	16.4	37.6	17.1	96.3
Organic N%	0.34	2.3	21.2	14.7	39.1	22.6	93.2
Carbohydrate C%	0.32	2.6	19.3	14.1	36.6	27.5	95.5
C/N	9.8	17.0	11.9	11.2	9.7	7.6	–

Soil organic matter is intimately associated in organomineral complexes and in particular with the clay fraction (Table 3.4). Free particulates of plant residues may be encrusted by adsorbed clay minerals. The residues become protected against further decay and form centers of aggregates (Fig. 3.1, 3.2). Investigations by scanning electron microscopy showed that many aggregates of 100 to 200 mm in diameter have cores of plant debris, and when such aggregates are disrupted by ultrasonic energy, fine fragments of partly degraded, but still structured plant particles can be obtained by density separation (Oades, 1993) (Fig. 3.3).

As decomposition proceeds production of binding agents in the form of hyphae, extracellular polysaccharides, strengthens the interaction with mineral particles and results in a stabilization of soil aggregates with occluded organic fragments. With time the more labile components are utilized by microorganisms, leaving organic residues that are more recalcitrant and less able to support the microbial population.

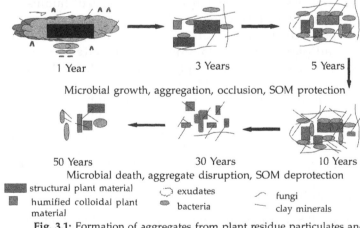

1 Year   3 Years   5 Years

Microbial growth, aggregation, occlusion, SOM protection

50 Years   30 Years   10 Years

Microbial death, aggregate disruption, SOM deprotection

■ structural plant material   ◌ exudates   ⌒ fungi
■ humified colloidal plant   ● bacteria   ⎯ clay minerals
   material

Fig. 3.1: Formation of aggregates from plant residue particulates and by association with clays.

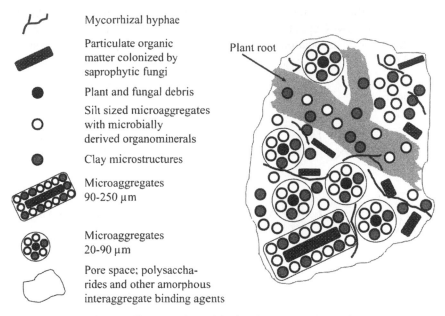

Mycorrhizal hyphae

Particulate organic matter colonized by saprophytic fungi

Plant and fungal debris

Silt sized microaggregates with microbially derived organominerals

Clay microstructures

Microaggregates 90-250 μm

Microaggregates 20-90 μm

Pore space; polysaccharides and other amorphous interaggregate binding agents

Fig. 3.2: Conceptual model of soil aggregate hierarchy.

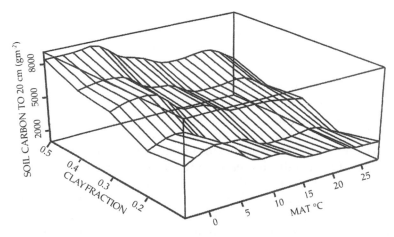

Fig. 3.3: Model prediction of soil carbon contents as function of clay content and mean annual temperature (MAT) (Burke et al., 1989; Parton et al., 1996).

Hassink (1997) reported a correlation between clay plus silt content and organic C and N contents. The protective effects of clay on soil C, N, P, and S have also been included to describe the turnover of soil organic C and associated nutrients (Fig. 3.3).

Golchin et al. (1994) observed large undecomposed root and plant fragments in the 500 μm to 2000 μm fraction and more decomposed materials in the 10 to 100 μm fractions. Cross Polarisation Magic Angle Spinning ^{13}C NMR spectroscopy provided insights into the chemical structure of SOM associated with different primary inorganic particle sizes (Fig. 3.4).

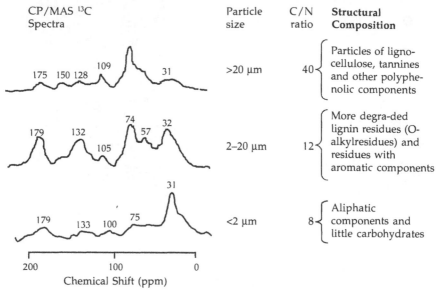

Fig. 3.4: Structural composition of the organic matter associated with decreasing particle size fractions (Baldock et al., 1992).

In general, anomeric O-alkyl resonances derived from carbohydrates or from ether bonds in degraded lignin (Fig. 3.5) predominate across the entire textural ranges, and are minimized in the silt (2 to 50 μm) range. Their strongest resonances are always observed in the sand fraction and resemble those of fresh plant litter. Aromatic C, and particularly that of the C-substituted aromatic C is maximized in silts but lowest in the clay fraction(s) < 2 μm. Alkyl C steadily increases with decreasing particle size, and represents in the clay fraction the major or second major C-species. As a reason for the strong alkyl signals a more significant contribution of natural waxes is considered (Oades et al., 1987).

Many studies show that particulate organic matter (POM) is closely linked to aggregation, and it is suggested that macroaggregates consist of microaggregates cemented together by plant residues in progressive stages of decay. Crushing of macroaggregates (>250 mm) provides particles of sand or quartz and of plant-derived material from roots or leaves colonized by microorganisms, which decompose them to form humified materials. The microorganisms form mucilages and attach to clay particles, so that the organic matter is enclosed in microaggregates (<250 mm), and thereby slowing its biodegradation. Even readily available compounds such as citric arid are stabilized in contact with clay minerals or mineral oxides, as shown in Fig. 3.6.

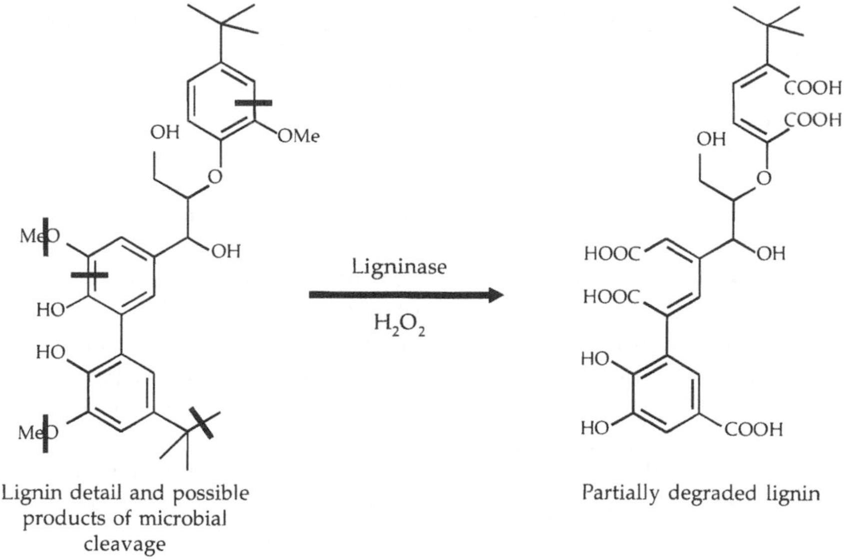

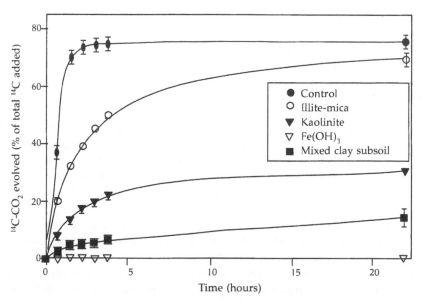

Lignin detail and possible products of microbial cleavage

Ligninase

H_2O_2

Partially degraded lignin

Fig. 3.5: Microbial cleavage of lignin structures.

Fig. 3.6: Inhibition of biodegradation of organic compounds resulting from adsorption to solid particles (Chenu and Stotzky, 2002).

Microscopic and electron microscopy evaluation gives direct access to the origin and the degree of decomposition of SOM within the different fractions. By using different strengths of ultrasonic energy to disrupt aggregates, the released light fraction (POM) can be differentiated into free inter-aggregate (between aggregates) and occluded intra-aggregate (inside of aggregates) POM (Baldock, 2002; Kölbl et al., 2004).

Accumulation of free organic particulates is favored in cold and dry climates and in continuously vegetated soils with a large return of plant litter, e.g. in grass or forest vegetation areas. Here they can account for 15 to 40% of the OM in surface horizons, whereas in tilled soils they contribute less than 10% of the OM, reflecting differences in cropping sequences and tillage.

It is estimated that through "recommended management practices (RMPs)" applied to the total area of U.S. land (916×10^6 ha) used by agriculture, grazing or forestry ecosystems the country could achieve sequestration of soil organic carbon at an annual rate 144–432 Tg C yr^{-1}, on average 288 Tg yr^{-1}. This amount would mitigate by 6.7 times the emission of 43 Tg C yr^{-1} by present agricultural activities.

Organic particles adhering to the surfaces of mineral particles form organomineral complexes. They can be recovered by minimal dispersion (e.g. density fractionation) of soil samples in which aggregates remain intact (Gregorich et al., 1997) and consist of partly decomposed litter particles from recently deposited crop residues or of older uncomplexed OM previously occluded in aggregates but released by their destruction. Organomineral complexes tend to be readily depleted when soils under a permanent native vegetation are brought into cultivation.

Many studies on different soil types showed that macroaggregates contained younger C than microaggregates. Examples from investigations in an Oxisol, Alfisol, and Inceptisol, in which C3 vegetation was replaced by C4 vegetation, are shown in Fig. 3.7.

Jastrow and Miller (1997) used the photosynthetic switch of C4 to C3 vegetation to calculate average turnover times of aggregate-associated organic matter. The average turnover time for old C4-derived C was 412 years for microaggregates compared with an average turnover time of 140 years for macroaggregates. Net input rates of new C3-derived C increased with aggregate size from 0.73 to 1.13 g kg^{-1} fraction per year.

Physical factors confer biological stability on organic matter in soils through the constraints they place for reactions between substrates and enzymes or decomposer organisms. Microbial access to substrates when they are located in small pores or sorbed on solid surfaces may be limited thereby protecting the OM from rapid decomposition. Carbohydrates and nitrogenous substances can interact with soil colloids by sorption, polymerization, or entrapment in voids of inorganic or organic soil constituents.

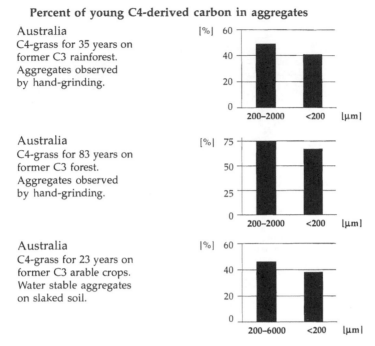

Percent of young C4-derived carbon in aggregates

Australia
C4-grass for 35 years on
former C3 rainforest.
Aggregates observed
by hand-grinding.

Australia
C4-grass for 83 years on
former C3 forest.
Aggregates observed
by hand-grinding.

Australia
C4-grass for 23 years on
former C3 arable crops.
Water stable aggregates
on slaked soil.

Fig. 3.7: Chemistry and nature of protected carbon in soil (Skjemstad et al., 1996).

It can be assumed that the observed continuous decline in bioavailability (and extractability) is associated with a slow diffusion of the molecules to more remote sites. Intraparticle and intraorganic diffusion into nanopores is believed to account for these slow phases of sorption and desorption of nonionic molecules. Even dissolved soil organic matter can diffuse into nanopores of primary particles or sorb on active surfaces. Within pores smaller than 8 nm this dissolved matter is inaccessible to exoenzymes. A conceptual model of this dissolved SOM stabilization in nanopores is shown in Fig. 3.8, according to Guggenberger and Haider (2002).

According to Amelung et al. (1998) organic matter in the coarse sand fraction is composed mainly of little decomposed plant residues, whereas that associated with fine sand consists of fairly altered organic debris and fine particles of roots. The CuO-oxidation method revealed that particle size fractions of soils differ in the contribution of lignin to total organic carbon, but also in the degree of lignin alteration. The highest lignin contents obtained by the CuO-oxidation method were noted in the coarse fractions and continuously decreased with decreasing particle size. Lignin in finer fractions had undergone more intensive microbial degradation revealed by a higher degree of oxidative alteration than in coarser fractions (Fig. 3.9).

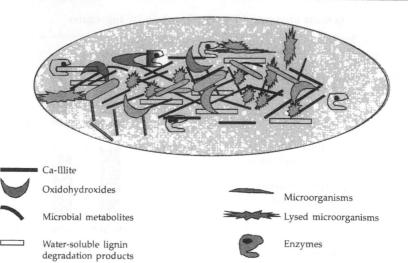

Ca-Illite

Oxidohydroxides

Microbial metabolites

Microorganisms

Lysed microorganisms

Water-soluble lignin
degradation products

Enzymes

Fig. 3.8: Model for the stabilization of SOM in the soil pore system (Guggenberger and Haider, 2002).

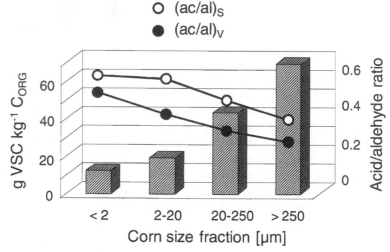

Fig. 3.9: Average contents of lignin related phenols (VSC) after CuO oxidation and acids to aldehyde ratios in different corn size fractions (Amelung et al., 1998).

Because of greater microbial alteration of the organic materials in finer fractions the C/N ratios were significantly lower than tthose in the coarser fractions, which was paralleled by higher ratios of microbe-derived sugars than plant-derived sugars.

^{13}C NMR spectroscopy is, at present, probably the most favored tool for detecting individual forms of C (Baldock et al., 1992). It has been used for over 30 years to study SOM compositions. More recent developments such as the use of higher magnetic fields, Fourier transform techniques, and cross-polarization magic-angle spinning (CP/MAS) have allowed the application to a wider range of samples and especially to whole soils with a minimum transformation of the samples. NMR spectroscopy provides important insights into the chemical structure of SOM associated with different primary particle fractions (Fig. 3.4).

Mahieu et al. (1999) published a statistical analysis of published CP/MAS ^{13}C-NMR spectra on 300 soils, and there was a remarkable similarity between all soils despite a wide range of land use, climate, cropping practices, and fertilizer or manure application. Functional groups in whole soils were always in the same abundance order with O-alkyls being most abundant, followed by alkyls (mean 25%), aromatics (mean 20%), and finally carbonyls (mean 10%). In general O-alkyl resonances derived from carbohydrates or from ether bonds in degraded lignin predominate across the entire textural range but are minimized in the silt fraction. The strongest O-alkyl signals are always observed in the sand fraction, and spectra of the organic matter associated with sand strongly resemble those of fresh plant litter. Aromatic C, and in particular the C-substituted aromatic C maximizes in silts, but is lowest in the clay fraction (Fig. 3.4).

Clearly distinguished are certainly the ^{13}C-NMR spectra of humic and fulvic acids with much higher contents of polar groups in the latter fraction. From these data Mahieu et al. (1999) reported that humic acids compared to whole soil contain more aromatics and fewer *O-alkyls*, which are the most abundant group in humic acids. The most striking point about fulvic acids is the high proportion of carbonyls (+15% compared to whole soils). The natural ^{13}C-labeling technique in which C3 vegetation was replaced by C4 vegetation, showed that particle-size fractions represent pools of SOM with different turnover times. Organic compounds in sand-size fractions are turned over rather rapidly within several years, whereas organic matter associated with particles below 50–20 μm show markedly higher turnover times, thus being involved in mid- and longterm dynamics of SOM.

Christensen and Bech-Andersen (1989) reported that amino acids, and in particular diaminopimelic acid, which is confined to the peptidoglycans of cell wall of the prokaryotes, are recovered in elevated yields from clay. Because most of the bacterial biomass has been found in the vicinity of clay, high concentrations of bacterial-derived compounds (with diaminopimelic and muramic acids as important bacterial cell wall compounds) in the clay-size separates they may be caused by a high production of microbial-derived substances.

Beside spatial stabilization of POM with structural relationship to plant tissue material at different stages of decomposition, the chemical stabilization of completely humified products, together with microbial compounds, makes an important contribution to SOM. Accumulation of refractory humic compounds can be caused by preferential sorption from refractory fractions of organic matter in the dissolved stage (dissolved organic matter, DOM). Their production occurs by leaching of dead plant residues particularly from those of the litter layer with water or by solubilization of already humified compounds in the soil solution. Only small proportions of this dissolved organic matter can be identified in the form of organic acids, sugars, amino acids, and phenols, but most DOM is complex humified organic matter.

Adsorption of DOM on mineral phases results in a drastic decrease of its microbial degradability. Laboratory studies showed that soil samples can adsorb dissolved organic carbon (DOC) rapidly, and evidence was provided for a fast and strong adsorption of DOM to Al- and Fe- oxides-hydroxides (Kalbitz et al., 2005; Chenu and Plante, 2006). Sorption of the hydrophobic DOM fraction was strongly favored over the binding of the hydrophilic fractions in soils and on hydrous oxides; ^{13}C-NMR spectroscopy showed that the sorption of DOM to soil material led to a preferential depletion of carbonyl and aromatic C, whereas alkyl C accumulated in the soluble portion. A preferential sorption of N compounds to oxides and clay minerals was also reported. Especially in the deeper layers of the forest floor and in the A-horizon low rates of DOM decomposition were found, which qualifies DOM as a source for the formation of stable humic substances. The main process by which DOM is retained in mineral horizons, and that adds to stabilized humus pools is likely to be sorption at the mineral horizons. Incubation studies of DOM from different origins showed that the mean residence time increased from 28 years in solution to 91 years after sorption. In forest soils from ^{13}C- and ^{14}C-data this sorption is estimated to equal 24 Mg C ha^{-1} and yr^{-1} (Kalbitz et al., 2005). Lignin contents regulate litter decomposition rates and are important for the production of DOM especially during later stages of litter decay.

The view that SOM stabilization is dominated by the selective preservation of recalcitrant organic compounds can no longer be accepted because the soil biotic community is able to disintegrate any plant derived-OM, including lignin (Kögel-Knabner et al., 2005).

Although most plant polysaccharides are readily decomposed, several microbially synthesized polysaccharides are relatively resistant to decomposition. Based on model experiments, mechanisms for their stabilization have been proposed; these include a close association with mineral colloids, or an interaction with metal ions through uronic acid and mannose units, or an interaction with tannins. The relatively stability of the typically as labile considered carbohydrate carbon may be explained by physical or chemical protection from degradation in the inorganic matrix or by association with or within humic molecules (Derrien et al., 2006). Furthermore, protection arises from the special role of polysaccharides as "bridges" between mineral particles, being directly involved in the formation of stable aggregates. There are indications that even more stable products of microbial synthesis are involved in the stabilization of SOM. One substance of importance is "glomalin", a glycoprotein with N-linked oligosaccharides, that is produced by hyphae of arbuscular mycorrhizal fungi, and which was reported to be abundant and stable in soil. Glomalin also promotes formation and stability of aggregates and thereby SOM contents (Preger et al., 2007).

Lignin, which is the second most abundant component of plant residues in terrestrial ecosystems, is relatively recalcitrant by itself, and degradation rates of lignocellulose materials are negatively correlated to their lignin content or to their lignin/N ratio (Shevchenko and Bailey, 1996). Lignin as an aromatic biomolecule is degraded at a much slower rate than cellulosic and non cellulosic polysaccharides and proteins. Reasons for this recalcitrance are the presence of non hydrolyzable C-O-C and C-C bonds between the phenylpropanoid units and the structural variety of stable ether and C-C bonds which can be cleaved only by oxidative mechanisms through lignolytic enzymes. These enzymes are involved in generation of lignin radicals by removing electrons from phenolic units. The resulting unstable radicals subsequently undergo a variety of spontaneous rearrangement reactions, and by adding water and oxygen molecules additional hydroxyl-, carboxyl- and carbonyl functions are introduced, which increase the oxygen content in the remnant lignin (Haider, 1992). Another peculiarity of lignin biodegradation, compared to other plant compounds, is the cometabolic character of the degradation process, by which it does not provide a source of energy or carbon for ligninolytic organisms. Thus, for an effective lignin degradation, readily degradable cosubstrates such as carbohydrates are required.

It was stated already that particle size fractions differ in their contribution of lignin, but also in the degree of lignin alterations (Fig. 3.4). Preservation of lignin constituents in humic substances has attracted attention for many years. Similarities and differences in structure and chemical reactivity between lignin and humic compounds were reviewed in terms of existing humification hypotheses (Kögel-Knabner, 1993).

Oxidative splitting with CuO or reductive splitting with hydroiodide, by direct pyrolysis or in combination with previous chemolysis by tetramethylammonium hydroxide revealed that phenolic structures characteristic for altered lignin structural units contribute to the SOM pool of soils after continuous farming for 15 years with corn (Chefetz et al., 2000).

Mechanisms for determining the stability of various SOM pools

Soils contain several tons of organic matter per hectare. Most of this SOM pool is only very slowly degraded by microorganisms. Models to describe the turnover of SOM assume the presence of easily and less easily degradable OM pools in soil, which exist in a sequence of overlapping residence times. Table 3.5 shows estimates of mean residence times for various fractions of SOM. The active or easily degradable pools with mean residence times of 0.1 to 4 years consist of free or only loosely adsorbed POM.

Table 3.5: Comparison of estimated mean residence times of soil organic matter in soil physical fractions (Guggenberger and Haider, 2002).

	Mean residence times (years)			
Pool	Jenkinson and Rayner (1977)	Parton et al. (1987)	Buyanovsky et al. (1994)	Carter (1996)
I	Decomposable plant material, 0.24	Metabolic plant residues, 0.1 to 1	Vegetative fragments (2 to 0.2 mm), 0.5 to 1	Litter, 1 to 3
II	Resistant plant material, 3.33	Structural plant residues, 1 to 5	Vegetative fragments (0.05 to 0.025 mm), 1 to 3	Free particulate material (light fraction), 1 to 15
III	Soil biomass, 2.44	Active SOM pool, 1 to 5	SOM in aggregates (2–1 mm), 1–4	Microbial biomass, 0.1 to 0.4
IV	Physically stabilized SOM, 72	Slow SOM pool, 25–50	SOM in aggregates (1–0.1 mm), 2 to 10	Intermicro-aggregate SOM[a], 5 to 50
V	Chemically stabilized SOM, 2857	Passive SOM pool, 1000 to 1500	SOM in fine silt, ca. 400 SOM in fine clay, ca. 1000	Intramicro-aggregate SOM[b]: Physically sequestered, 50 to 1000 Chemically sequestered, 1000 to 3000

(a) Organic matter stored within macroaggregates but external to microaggregates; includes coarse occluded particulate organic and microbially derived organic matter.
(b) Organic matter stored within microaggregates; includes fine occluded particulate organic matter and microbially derived organic matter.

Whereas the compounds of the active and some of the slow pools are used by microorganisms for metabolism and energy production, the great portion of humic compounds in the passive pool is not readily available as a source of energy and microbial biosynthesis, because they represent refractory plant and microbial residues. Their diversity and lack of regular polymeric structures require a large number of enzymatic steps to degrade them to CO_2 and to use them for energy production. Their degradation, however, accelerates with small inputs of readily available substrates (De Nobili et al., 2001). Even when these inputs are relatively small, they "trigger" the synthesis of enzymes for the cometabolic degradation of a portion of the passive SOM fraction. This priming effect lasts only for a relatively short period after addition and disappears after metabolization of the added substrates, but it can be hypothesized that this effect limits the accumulation of the passive pool. A similar conception of how this could occur was similarly presented by Fontaine et al. (2003) and is shown in Fig. 3.10.

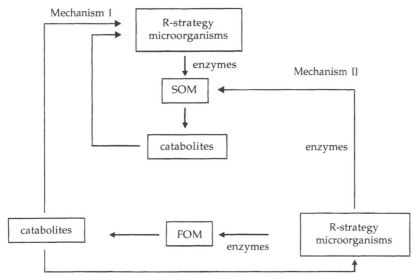

Fig. 3.10: Two mechanisms to explain the degradation of the passive SOM pool (Fontaine et al., 2003).

The most active organisms known to degrade humic compounds are basidiomycetes and actinomycetes, and the most active organisms were *Phanerochaete chrysosporium*, and *Trametes versicolor* (Haider and Martin, 1988; Kästner, 1995), which are also known as active degraders of lignin by producing peroxidases. Both organisms were shown to generate radicals during the bleaching process of isolated humic acids. This priming effect, which is caused by small amounts of easily available plant residues or compounds (fresh organic matter, FOM), fades after a short period, and a repeated addition is needed to enhance the degradation of the passive organic matter pool and limits its accumulation (De Nobili et al., 2001).

REFERENCES

Amelung W, Zech W, Zhang X, Follett H, Thiessen H, Knox E, Flach K-W, 1998: Carbon, nitrogen, and sulfur pools in particle size fractions as influenced by climate. Soil Sci. Soc. Am. J. 62:172–181.

Amelung W, Zech W, Zhang XD, Saijapongse A, Niamskul C, 1998: Lignin and carbohydrates in soils under secondary forest, alley-cropping, and continuous farming, Thailand. Z. Pflanzenern. Bodenk. 161, 297–302.

Anderson DW, Saggar S, Bettany JR, Stewart, JWB, 1981: Particle size fractions and their use in studies of soil organic matter. I. The nature and distribution of forms of carbon, nitrogen and sulfur. Soil Sci. Soc. Am. J. 45, 767–787.

Baldock JA, Oades AG, Waters X, Peng AM, Vasallo AM, Wilson MA, 1992: Aspects of the chemical structure of soil organic materials as revealed by solid-state 13C NMR spectroscopy. Biogeochemisty 16: 1–42.

Baldock JA, 2002: Interactions of organic materials and microorganisms with minerals in the stabilization of strucrure. In: Effect of mineral colloids on biogeochemical cycling of C, N, P and S in soil. IUPAC Series on Analytical and Physical Chemistry, Vol. 8, 85–132.

Batjes NH, 1996: Total carbon and nitrogen in the soils of the world. Europ. J. Soil Sci. 47: 151–163.

Bettany JR, Stewart JWB, 1981: Particle size fractions and their use in studies of soil organic matter: I. The nature and distribution of forms of carbon, nitrogen, and sulfur. Soil Sci. Soc. Am. J. 45:767–772.

Burke IC, Yonker CM, Parton WJ, Cole CV, Flach K, Schimel DS, 1989: Texture, climatic and cultivation effects on soil organic matter content in US grassland soils. Soil Sci. Soc. Amer. J. 53, 800–805.

Buyanovsky GA, Aslam M, Wagner GH, 1994: Carbon turnover in soil physical fractions. Soil Sci. Soc. Am. J. 58, 1167–1187.

Carter MR, 1996: Analysis of soil organic matter in agroecosystems. In: Structure and organic matter storage in agricultural soils. M.R. Carter, B.A. Stewart (eds.). Adv. Soil Sci. CRC Press, Boca Raton FL.

Chefetz B, Chen Y, Clapp E, Hatcher PG, 2000. Characterization of organic matter in soils by thermochemolysis using tetramethylammonium hydroxide (TMAH). Soil Sci. Soc. Am. J. 64, 583–589.

Chenu C, Plante AF, 2006: Clay sized organo-mineral complexes in a cultivation chronosequence: revisiting the concept of the "primary organo-mineral complex". European J. Soil Sci. 57, 596–607.

Chenu C, Stotzky G, 2002: Interactions between soil particles and microorganisms: an overview. In: Interactions between soil particles and microorganisms. IUPAC Series on Analytical and Physical Chemistry, Vol. 8, Ch. 7, 3–40. John Wiley and Sons. Ltd., New York.

Christensen BT, 1996: Carbon in primary and secondary organomineral complexes. In: Structure and Organic Matter Storage in Agricultural Soils. M.R. Carter, B.A. Stewart (eds.). CRC-Press Inc., Boca Raton FL.

Christensen BT, Bech-Andersen S, 1989: Influence of straw disposal on distribution of amino acids in soil particle size fractions. Soil Biol. Biochem. 21, 35–40.

De Nobili M, Contin M, Mondini M, Brookes PC, 2001: Soil microbial biomass is triggered intoactivity by trace amounts of substrates. Soil Biol. Biochem 33, 1163–1170.

Derrien D, Marol C, Balaban M, Balesdent J, 2006: The turnover of carbohydrate carbon in a cultivated soil estimated by 13C natural abundance. Europ. J. Soil Sci. 57, 547–557.

Dignac M-F, Bahri H, Rumpel C, Rasse DP, Bardoux J, Balesdent J, Girardin, Chenu C, Mariottii A, 2005: Carbon–13 natural abundance as a tool to study the dynamics of lignin monomers in soil: an appraisal at the Closeaux experimental field (France). Geoderma 128: 3–17.

Dobermann A, Cassman KG, 2004: Improving the productivity. Plant Soil 247; 25–39.

Fontaine S, Mariotti A, Abadie L, 2003: The priming effect of organic matter: a question of microbial competition? Soil Biol. Biochem. 35, 837–843.

Germida JJ, Siciliano SD, 2000: Phosphorus, sulfur and metal transformations. In: Sumner M (ed). Handbook of Soil Sci. pp. 95 –106. Boca Raton Fl, CRC Press.

Golchin A, Oades JM, Skjemstad, Clarke P, 1994: Soil structure and carbon cycling. Austr. J Soil Res. 32, 1043–1064.

Gregorich EG, Drury CF, Ellert BH, Liang BC, 1997: Fertilization effects on physically protected light fraction organic matter. Soil Sci. Soc.Am. J. 61, 482–484.

Guggenberger G, Haider K, 2002: Effect of mineral colloids on biogeochemical cycling of C, N, P and S in soil. In: Interactions between soil particles and microorganisms.

IUPAC Series on Analytical and Physical Chemistry, Vol. 8, Ch 7, 267–322. John Wiley and Sons. Ltd., New York.

Guggenberger G, Kaiser K, Zech W, 1998: Mobilization and immobilization of dissolved organic matter in forest soils. Z. Pflanzenernähr. Bodenkd. 181, 401–408.

Haider K, 1992: Problems related to the humification processes in soils of temperate climates. In: G. Stotzky and J.M. Bollag, eds., Soil Biochemistry, Vol. 7, M. Dekker, New York S. 55–94.

Haider K, Martin JP, 1988: Mineralization of ^{14}C-labeled humic acids and humic-acid bound 14C-xenobiotics by *Phanerochaete chrysosporium*. Soil Biol. Biochem. 20, 425–429.

Hanson PJ, Edwards NT, Garten CT, Andrews GA, 2000: Separating root and soil microbial contribution to soil respiration. A review of methods and observations. Biogeochem. 48, 114–146.

Hassink J, 1997: The capacity of soils to preserve organic C and N by their association with clay and silt particles. Plant Soil 191: 77–87.

Hatcher PG, Spiker EC, 1988: Selective degradation of plant biomolecules. In: Humic Substances and Their Role in the Environment, F.H. Frimmel and R.F. Christman, eds., Dahlem Workshop, Berlin 1987, J. Wiley and Sons, Chichester, pp. 59–74.

Hattaka A, 1994: Ligninolytic enzymes from selected white-rot fungi: production and role in lignin degradation. FEMS Microbiol. Rev. 13, 125–135.

Hayes MHB, McCarthy P, Malcolm RI, Swift RS, 1989: Humic Substances: The search of structure of humic compounds. Wiley and Sons, Chichester. 764 pp.

Hedges JI, 1988: Polymerization of humic substances in natural environments. In: Humic Substances and Their Role in the Environment, F.H. Frimmel and R.F. Christman, eds., Dahlem Workshop, Berlin 1987, J. Wiley and Sons, Chichester S. 45–58.

Jastrow JD, Miller RM, 1997: Soil aggregate stabilization and carbons questration: feedbacks through organomineral associations. In: Soil processes and the carbon cycle. R. Lal et al. (ed.) CRC Press, Boca Raton FL. 207 pp.

Jenkinson DS, Rayner JH, 1977: The turnover of soil organic matter in some of the Rothamsted classical experiments. Soil Sci. 123, 298–305.

Jenkinson DS, 1977: Studies on the decomposition of plant materials in soil. The effect of plant cover and soil type on the loss of carbon from 14C-labeled ryegrass decomposing under field coditions. J. Soil Sci. 28, 417–423.

Jones DL, Edwards AC, 1998: Influence of sorption on the biological utilization of two simple carbon substrates. Soil. Biol. Biochem. 30, 1895–2006.

Kaiser K, Guggenberger G, 2003: Mineral surfaces and soil organic matter. Eur. J. Soil Sci. 54, 219–236.

Kaiser KJ, Bargholz J, Dardenne P, 2006: Lignin degradation controls the production of dissolved organic matter in decomposing foliar litter. Europ. J. Soil Sci. 57, 504–516.

Kalbitz K, Schwesig D, Rethemeyer J, Matzner E, 2005: Stabilization of dissolved organic matter by sorption to the mineral soil. Soil Biol. Biochem. 37, 319–1331.

Kästner M, 2000: The "humification" process or the formation of refractory soil organic matter. Biotechnology Vol. 11 b. Environmental Processes II, Chap. 4, p. 90–125, Wiley VCH, Weinheim FRG.

Kinchesh P, Powlson DS, Randall EW, 1995: ^{13}C NMR studies of organic matter in whole soils: I. Quantitation possibilities. Europ. J. Soil Sci. 46, 125–138.

Kingery WL, Simpson AJ, Hayes MBH, Locke MA, Hicks RP, 2000: The application of multidimensional NMR to the study of soil humic substances. Soil Sci. 165, 481–494.

Kirschbaum MUF, 2006: The temperature dependence of organic-matter decomposition— still a topic of debate. Soil Biol. Biochem. 38, 2510–2518.

Kögel-Knabner I, 1993: Biodegradation and humification processes in forest soils. In: Soil Biochemistry Vol. 8, J.M. Bollag and G. Stotzky, eds., M. Dekker Inc., New York, pp. 101–137.

Kögel-Knabner I, von Lützow M, Flessa H, Guggenberger G, Marschner B, Matzner E, Ekschmitt K, 2005: Mechanism and regulation of organic matter stabilization in soils. Geoderma 124, 176 pp.

Kögel-Knabner I, Ekschmitt K, Matzner E, Guggenberger G, 2006: Stabilization of organic matter in temperate soil conditions – a review. Europ. J. Soil Sci. 57, 426–445.

Kölbl A, Leifel DJ, Kögel-Knabner I, 2004: A comparison of two methods for the isolation of free and occluded particulate organic matter. J. Plant Nutr. Soil Sci. 168, 660–667.

Ladd JN, Foster RC, Nannipieri P, Oades JM, 1995: Soil structure and biological activity. In: Soil Biochemistry Vol. 9, G. Stotzky and J.M. Bollag eds., M. Dekker Inc., New York,

Ladd JN, Foster RC, Skjemstad JO, 1993: Soil structure: Carbon and nitrogen metabolism. Geoderma 56: 401–434.

Lal R, 2003: Management impact on compaction in forest soils. In: The potential of US Forest Soils to sequester carbon and mitigate the greenhouse effect. J.M. Kimble et al. (eds.) CRC/Lewis Publ., Boca Raton, FL, pp. 239–256.

Lal R, Follett RF, Kimble JM, 2003: Achieving soil carbon sequestration in the UNITED States. A challenge to the policy makers. Soil Sci. 168:827–845.

Mahieu N, Powlson DS, Randall EW, 1999: Statistical analysis of published Carbon–13 CPMAS NMR spectra of soil organic matter. Soil Sci. Soc. Am. J. 63, 307–319.

Martin JP, Haider K, 1986: Influence of mineral colloids on turnover rates of soil organic carbon. In: Interactions of soil minerals with natural organics and minerals. P.M. Huang and M. Schnitzer (eds.). SSSA Special Publication Number 17, pp. 283–304.

Oades JM, 1993: The role of biology in the formation, stabilization and degradation of soil structure. Geoderma 56:377–400.

Oades JM, Vassallo AM, Waters AG, Wilson MA, 1987: Characterization of organic matter in particle size and density fractions of a red-brown earth by solid state ^{13}C NMR. Austr. J. Soil Res. 25, 71–82.

Parton WJ, Mosier AR, Ojima DS Valentine DW, Schimel DS, Weier K. Kulmala AE, 1996: Generalized model for N_2 and N_2O production from nitrification and denitrification. Global biochem. Cycles. 10, 401–412.

Parton, WJ, Schimel DS, Cole CV, Ojima DS, 1987: Analysis of factors controlling soil organic matter level in Great Plains grasslands. Soil Sci. Soc. Am. J. 51:1173–1179.

Preger AC, Rilling MC, Johns AR, Du Preez CC, Lobe I, Amelung W, 2007: Losses of glomalin-related protein under prolonged arable cropping: A chronosequence study in sandy soils of the South African Highveld. Soil Biol. Biochem. 39, 445–453.

Robert M, Chenu C,1992: Interactions between soil minerals and microorganisms. In: Soil Biochemistry Vol. 7, G. Stotzky und J.M. Bollag (eds.), M. Dekker, Inc. New York, S. 307–404.

Sauerbeck DR, Gonzalez MA, 1977: Field decomposition of carbon–^{14}- labeled plant residues in various soils of the Federal Republic of Germany and Costa Rica. Proc Symp. IAEA and FAO. Vol 1: p. 117–132.

Schlesinger WH, Andrews JA, 2000: Soil respiration and the global carbon cycle. Biogeochem. 48: 7–20.

Schlesinger WH, 1991: The global carbon cycle. In: Biogeochemistry – An Analysis of global change. p. 308–321. Academic Press, San Diego, California.

Schlesinger WH, 1997: Biogeochemistry: An analysis of global change. Academic Press, San Diego.

Shevchenko SM, Bailey GW, 1996: Life after death: Lignin-humic relationships reexamined. Critical Rev. Environ. Sci. Technol. 26: 95–153.

Simpson AJ, Kingery WL, Hatcher PG, 2003: The identification of plant derived structures in humic materials using three-dimensional NMR spectroscopy. Environ. Sci. Technol. 37, 337–342.

Skjemstad JO, Clarke P, Taylor JA, Oades JM, McClure SG, 1996: The chemistry and nature of protected carbon in soil. Austr. J. Soil Res. 34, 251–271.

Wander MM, Bidart MG, 2000: Tillage practice influences on the physical protection, particulate organic matter. Biol. Fert. Soils 32, 360–367.

4

Turnover of Nitrogen, Phosphorus and Sulfur in Soils and Sediments

Soil organic matter (SOM) contains the elements N, P, and S in organic linkages (Table 4.1). To be available to microorganisms or plants these linkages have to be transformed into the mineral forms. Therefore, their mineralization limits their availability as nutrients. Because their mineralization process is not fast enough to cover the demand of plants, in particular of those used in human and animal nutrition, it is necessary to subsidize their growth by additional inorganic and/or organic fertilization.

Table 4.1: Contents of N, P, and S in soil distributed between inorganic and organic fractions.

Element ratios in soil	N-content in European soils	P-content in European soils	S-content in European soils
C/N 10 to 15 and above	5 to 8 t N ha^{-1} bound in humic matter of which 80 to 120 kg are contained in microbial biomass	–	–
C/P 50 to100	–	2 to 25 t P ha^{-1} bound of which 60 to 65% is organically bound and 35 to 40% inorganically bound (mainly to Fe and Ca minerals)	–
C/S 60 to 200	–	–	1.7 to 30 t S ha^{-1} of which 3 to 10% is inorganically bound, 20 to 30% is inorganically bound and 40 to 60% is bound in form of sulfate esters

Turnover of these nutrients into plant-available forms also occurs in the atmosphere and water bodies as well; these processes have far-reaching effects upon other systems including water, the atmosphere, and civilization.

Relations of C, N, P, and S in SOM

The relations of C, N, P, and S together are more or less fixed, although they vary within some limits. C/N ratios fluctuate between 10 to 15:1, C/P vary between 50 to 100:1, and C/S between 60 to 200:1. According to Roberts et al. (1989) the C:N:P:S ratios in Canadian soils varied between 68:6.9:1.4:1 and 145:11.6:1.8:1. Because transformation of N, P, and S in mineral forms is generally connected with the mineralization of carbon, the geochemical cycles of all these elements are tightly connected (Tate, 1987).

Many reactions in the N and S cycles are combined with aerobic or anaerobic conditions and accompanied by changes of valences (Fig. 4.1, 4.2, 4.3).

Reactions	Redox potential [mV]	Chemistry	Microorganisms
Nitrate reduction Denitrification Nitrate respiration	+600 to +400	$2NO_3^- + 2\,H^+ + 10\,H^+$ → $2\,N_2 + 6\,H_2O$	*Pseudomonas* spp. *Thiobacillus* spp. *Alcaligenes*
$Mn^{4-} \rightarrow Mn^{3-}$ $Fe^{3-} \rightarrow Fe^{2-}$	+600 to +400 + 600 to +300	MnO_2 → $Mn^{2-} + 2\,H_2O$ $Fe_2O_3 + n\,H_2O$ → $Fe^{2-} + H_2O$	Microorganisms that reduce Fe^{3-} often reduce also Mn^{4-}
$SO_4^{2-} \rightarrow H_2S$ Dissimilatory sulfate reduction Sulfate respiration	+ 100 to –200	$SO_4^{2-} + 2\,H^+ + 8\,H^+$ → $H_2S + 4\,H_2O$	*Desulfovibrio* spp. *Desulfomonas* spp. *Desulfo maculum* etc.
Formation of CH_4 Methanogenesis	–150 to –200	CH_3COOH → $H_2 + CO_2$ → $CH_4 + 2\,H_2O$ Other Cl products	*Methanobacterium* spp. *Methanococcus* spp. *Methanosarcina* spp.

Fig. 4.1: Changes of valencies in the N and S cycles.

Properties and forms of nitrogen in soil

^{13}C-NMR spectroscopy has contributed significantly to our knowledge of the chemical structure of SOM. The application of the solid-state-high-resolution-CP-MAS-technique even permits the study of complete soils and native humic materials without any prior chemical treatment, such as, for instance the extraction by aqueous sodium hydroxide (Fründ and Lüdemann, 1989).

Ratio of C/N, C/P, C/S in soils	N content in European soils	P content in European soils	S content in European soils
C/N 10–15 C/P 50–100 C/S 60–200	5–8000 kg N ha^{-1} bound to humic matter with 80–120 kg bound in microbial biomass	2000–2500 kg P ha^{-1} with 60–65% bound to organic matter and 30–40% to inorganic matter (mostly to Fe- and Ca-structures)	1700–30000 kg S ha^{-1} with 20–30% bound to organic matter and 3–10% to inorganic matter; 40–60% bound as sulfate ester

Fig. 4.2: Contents of N, P and S in European Soils and distribution to inorganic and organic soil fractions.

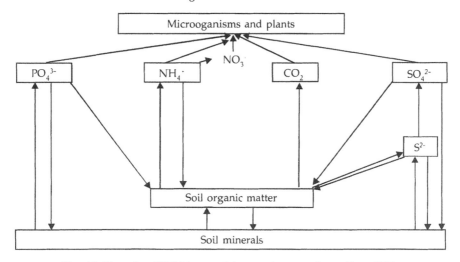

Fig. 4.3: The role of SOM in supplying nutrients to plants (Tate, 1987).

^{15}N-NMR at natural abundance encounters greater sensitivity problems caused by the low natural abundance this isotope (only 0.27%) and its small and negative gyromagnetic ratio. Consequently, no serious attempts have been made to determine the chemical structure of SOM nitrogen by CP-MAS-technique in spite of the central importance of SOM nitrogen in humic and fulvic acid materials. Attempts have been made to learn by application of the CP-MAS-technique to biodegraded plants grown on ^{15}N-highly enriched fertilizer. The spectral changes observed are compared to CP-MAS ^{13}C-NMR spectra of the same composts.

The spectrum (in Fig. 4.4) shows the unfermented starting material and in Fig. 4.5 in the ^{15}N spectrum additional signals are found at 444, 139, and 114 ppm and weak shoulders at 87 , 52, 40 ppm. Spinning side bands (first and second order) are marked with asteriks or arrows. The signal centered on 87 ppm represents 82 to 87% of the total intensity and is assigned to the peptide nitrogens of proteins. The most plausible assignments of the other resonances are shown in Table 4.2.

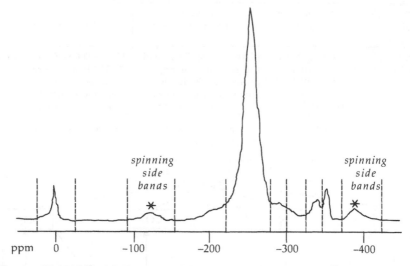

Fig. 4.4: CP-MAS ^{15}N-NMR-spectra of the incubated plant residues (^{15}N–labelled wheat, incubated for 631 days at 60% water holding capacity, [WHC]) (Knicker et al., 1997).

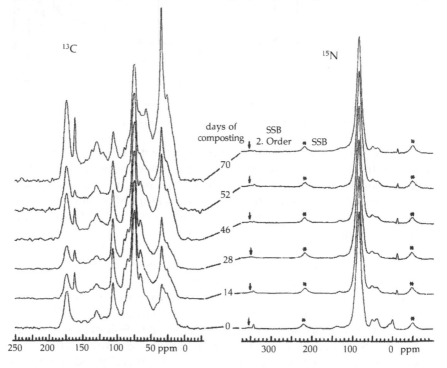

Fig. 4.5: CP-MAS ^{15}N-NMR- and ^{13}C-NMR spectra of freeze dried compost L. rigidum plant material (Almendros et al., 1995).

Table 4.2: Possible assignments of the ^{15}N-NMR spectra of the ^{15}N-labeled composts shown in Fig. 4.4 (Knicker et al., 1997).

Signal (ppm)	Rel.- intensity (%)		Assignment
	Initial (I)	Final (F)	
344	2	<1	Residual nitrate ions
144 to 140	4	5	Nitrogen at position 9 in purine base of nucleic acids
87	82	87	Peptide nitrogen of protein
40 to 55	8	5	-NH$_2$ in bases of nucleic acids
0	4	<1	Free amino groups of peptides, amino acids, and amino sugars
–10	0	4	NH$_4$$^+$
Missing signals			
950	–		Quinonimines
160 to 180	–		Pyrroles
280 to 320	–		Schiffbases
Around 290	–		Phenazine derivatives

During composting the signals from the free amino groups and the nucleic acids are diminished most rapidly. Simultaneously NH$_4$$^+$ (–9 ppm) is formed, but no new signals appear. It is therefore improbable that even a long-term composting process leads to the formation of new heteroaromatic nitrogen-containing rings or Schiff bases, as claimed by Anderson et al. (1989) and by Schulten et al. (1995).

^{13}C-NMR signals have been assigned to the presence of various functional groups in SOM (Table 4.3).

Table 4.3: Assignment for the ^{13}C-NMR spectra of SOM referenced to tetramethylsilane (Malcolm 1989).

Chemical shift range (ppm)	Assignment
160 to 120	Carboxyl/carbonyl/amide carbons
160 to 140	Aromatic COR or CNR groups
140 to 110	Aromatic CH carbons, alkene carbons
110 to 90	Anomeric carbons of carbohydrates, C2, C6 of syringyl units
90 to 60	Carbohydrate-derived structures (C2 to C5) in hexoses, Cα of some amino acids, higher alcohols
60 to 45	Methoxyl groups and C6 of carbohydrates and sugars, Cα of most amino acids
45 to 0	Methylene groups in aliphatic rings and chains, terminal methyl groups

Turnover of nitrogen in the environment

Nitrogen available to plants is often most limiting for plant growth. The great source of atmospheric dinitrogen (N$_2$) is relatively inert and can only be used by free living and symbiotic prokaryotic bacteria with the capacity for N_2-fixation. Considerable energy (946 kJ mol^{-1}) is required to break the triple bond between the N_2 molecules. N in the prehuman world resulted from biological N-fixation and this was the dominant way N was made available to living organisms, but with the advent of humans the amount that circulated naturally among various compartments of the biosphere was too small to cover the demand of the rapidly growing world population.

Since 1913, with the introduction of the Haber-Bosch process the production of fertilizer N tremendously increased and has been paralleled by a population increase (Fig. 4.6). Today, more than half of the food eaten by the world's population is nourished using N-fertilizer from the Haber-Bosch process (Evans, 1998).

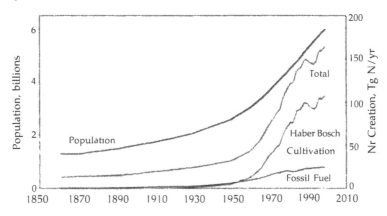

Fig. 4.6: Global development of reactive N production and population from 1860 to 2000. This figure shows the rapid increase of fertilizer N production together with population increase during the last 50 years and the global production and distribution of N available to organisms.

Pathways of nitrogen loss and their environmental impacts

The global reactive N supply increased from 1860 to the early 1990s by a factor of about 1.5 and is estimated to increase further 1.3 times until 2050 (Table 4.4).

Before 1900 creation of reactive nitrogen was dominated by biological N-fixation in natural ecosystems. At present synthetic fertilizer N contributes about 28 Tg N y^{-1} whereas managed biological fixation adds 20 Tg yr^{-1} of N and recycling of organic wastes between 28 and 36 Tg yr^{-1}. However, only half of all these anthropogenic N inputs on croplands are taken up by crops and their residues (Dobermann and Cassman, 2004).

Unfortunately, often less than 50 to 60% of the N applied to crops is recovered by plants (Fig. 4.7). The residue of applied N is easily transformed to various oxidized and reduced forms and readily distributed by hydrologic and atmospheric transport processes and can be lost through soil erosion, runoff, or leaching of nitrate or through gaseous emissions to the atmosphere in the form of ammonia, nitrogen oxides (NO and NO_2) and nitrous oxide (N_2O) or N_2. All these avenues of loss have environmental impacts or have implications for human health (Kroeze et al., 1999).

Table 4.4: Global reactive N creation and distribution beween 1860 and 2050 (estimations) in Tg N yr^{-1} (data from Galloway et al., 2004).

N creation (Tg yr^{-1})	1860	Early 1990	2050
Natural by lightning	5.4	5.4	5.4
Biol. N-fixation (terrestrial)[a]	120	107	98[a]
Biol. N-fixation (marine)	121	121	121
Anthropogenic Haber-Bosch	0	100	165
Production by fossil fuel combustion	0.3	24.5	52.2
NH_3 terrestrial emission	14.9	52.6	113
NH_3 marine emission	5.6	5.6	5.6
N_2O terrestrial emission	8.1	10.9	13.1
N_2O marine emission	5.6	5.6	5.6

(a) Estimates of biological N-fixation have many uncertainties because they are frequently carried out in areas where cultivated biological N-fixing plants make an important contribution to the N-fixing species.

One important question considers whether losses of N from synthetic mineral sources differ from those of organic origin, such as composts or manure. However, only a limited number of comparisons are available. Table 4.2 shows examples of the fate of nitrogen after applying ^{15}N-labelled fertilizer or legume residues. The range of estimated losses from both sources seems to be rather similar. It is, however, not clear how many of the comparative studies shown in Table 4.5 have used best management systems, because the legume inputs do not represent only shoot materials, which ignores the potentially large contribution associated with legume roots and nodules (Rochester et al., 2001; Peoples, 2004).

Transformations of ammonia N in soils by microorganisms are summarized in Table 4.6.

Table 4.7 shows that regardless of whether N is applied as fertilizer- or legume-N about 50% or less of this applied N remains unrecovered in crops or soil.

Table 4.5: Transformation of ammonia-N in soil (Schlesinger and Hartley, 1992).

Source	Mean 10^6 t N yr^{-1}	Range 10^6 t N yr^{-1}
Domestic animals	32	24 to 40
Sea surface	13	4 to 18
Forest and savannah soils	10	6 to 45
Fertilized soils	9	5 to 10
Combustion of biomass	5	1 to 9
Human excretions	4	
Coal combustion	2	
Traffic	0.2	
Total sources	75	50 to 128
Sink		
Wet deposits	30	
Dry deposits	10	
Sea surface	16	
Reaction with atmospheric OH radicals	1	
Total sinks	57	

Table 4.6: Mechanisms of ammonia transformations in soils.

1. Ammonia is absorbed and fixed by microorganisms.
2. Ammonia is absorbed by plants but it is also fixed in soil on clay minerals. This fixed ammonia can be released by Ca^{2-} or $K^{.}$ from this exchange complex.
3. By microbial and abiotic sorptive reactions, ammonia is fixed in SOM and becomes stabilized for long periods.
4. $NH_4^{.}$ is used by autotrophic bacteria (nitrifying organisms) as a source of energy and transformed into nitrate. Nitrate is also formed by heterotrophic organisms.
5. Ammonia released from nitrogen fertilizers and from urea in liquid manure is volatilized into the atmosphere (Table 4.4).

Table 4.7: N uptake and recovery in plants and soil (Rochester et al., 2001; Peoples, 2004).

Source of N applied	Crop uptake (% of appl. N)	Recovery in Soil (% of appl. N)	Total recovered in crop and soil (% of appl. N)	Unrecovered (lost) (% of appl. N)
Rainfed cereal cropping				
Fertilizer	16 to 51	19 to 38	54 to 84	16 to 46
Legume	9 to 19	58 to 83	64 to85	15 to 36
Irrigated cotton				
Fertilizer	–	–	4 to 17	83 to 96
Legume	–	–	62 to 82	18 to 38
Lowland rice				
Fertilizer	–	–	61 to 65	35 to 39
# Legume	–	–	87 to 93	7 to 13

Estimates of emissions of gaseous N products do not imply key differences in N losses between fertilizer N and manure sources, but measurements of the nitrate amounts leached from grazed pastures and other land suggest that the amounts lost are more a function of the size of annual input N (Fig. 4.7).

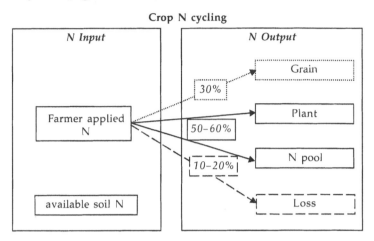

Fig. 4.7: Nitrogen use efficiency (Burke et al., 2005).

Nitrogen use efficiency can be increased substantially through a combination of management techniques and ranges from 20 to 50% in major cropping systems, but even excellent labor intensive management never reaches beyond 60 to 80% efficiency (Dobermann and Cassman, 2004).

Nitrification and nitrate leaching

Nitrification is the oxidation of NH_3 and other reduced forms of N ultimately to nitrate. The relatively immobile ammonia is transformed by this process into nitrate which is more easily leached and transported into ground- and surface waters.

Organisms capable of oxidizing NH_4^+ to NO_2^- and further to NO_3^- were first isolated by Winogradsky (1890). He isolated two distinct types of autotrophic bacteria (chemoautotrophs) obtaining their energy from the oxidation of NH_4^+ to NO_2^- in two steps and further to NO_3^-. These nitrifiers assimilate CO_2 via the Calvin cycle to form their cellular components. The first step is catalyzed by a monooxygenase that transfers one atom of O_2 to NH_3 from N-fertilizer.

Since nitrate leads to contamination of ground and surface water, different available chemicals are available to inhibit the formation of nitrate (Table 4.8).

Table 4.8: Chemical inhibitors of nitrification in soil.

Chemical	Formula	Application	Comment
Dicyanodiamide Didyn or DCD	$(CN-NH_2)_2$	Used in Germany as a nitrification inhibitor	Acts also as an N fertilizer
Ammoniumthiosulfate	$(NH_4)_2S_2O_3$	Acts as an nitrification inhibitor at low concentrations	Common use as a soil sterilizer
Nitrapyrine (N-serve)	2-chloro–6- (trichloromethyl- pyridine)	Most common nitrification inhibitor in the U.S.A.	Additive to liquid ammonia or urea solutions
Dimethylpyrazolphosphate (DMPP)	3,4- dimethyl- pyrazol-phosphate	New nitrification inhibitor	
Acetylene and derivatives	C_2H_2, phenyl- acetylene $C_6H_5C_2H_2$	Inhibits nitrifi- cation at very low concentrations	Used also to follow denitrification and N-fixation (see sections 4.4 and 4–4–2)

Biological nitrogen fixation

Many diverse eubacteria out of 27 families (Table 4.4) can form NH_3 from N_2. These prokaryotic organisms use ATP as energy to initiate the bond-breaking reaction: $N \equiv N + 3 H_2 \rightarrow 2 NH_3$. Theoretically as much as 28 moles of ATP are consumed in the reduction of 1 mole of N_2.

Although connected with uncertainties global biological nitrogen fixation is on the order of 90 to 130 Tg N yr^{-1}. Recently, however, it has been surpassed by the consumption of N from industrial processes (see Table 4.9). In the terrestrial habitat, the symbiotic fixation of nitrogen by rhizobia accounts for the largest contribution of combined nitrogen and is two to three times higher than rates exhibited by free-living nitrogen-fixing bacteria.

Table 4.9: Global estimates of biological N-fixation in Tg N yr^{-1} (Quispel, 1974; Burns and Hardy, 1975).

Ecosystem	Area (ha)	Tg N yr^{-1}
Leguminous plants	250 x 10^6	14 to 35
Rice	135 x 10^6	4
N-fixing diazotrophic bacteria in sugar cane and prairie grasses	3,000	45
Forests and woodland	4,100	40
Total solid land including free-living bacteria	36,100	118 to 139

Biological N_2-fixation can be conducted by free-living bacteria or those in symbiosis with higher plants. These are heterotrophic aerobic and anaerobic bacteria as well as phototrophic cyanobacteria (blue-green algae).

The symbiotic organisms exist in roots, leaves or stem galls and in the life community of fungi and cyanobacteria in lichens. These free-living organisms are widely distributed in soils (Table 4.10).

Table 4.10: N_2-binding bacteria in soils and other environments (from Haider, 1996).

Organism	Order	Association with plants	Occurrence
Free-living bacteria			
Aerobic	Azotobacter	–	In slightly acid or neutral soils
Facultative aerobic	Klebsiella	–	
	Enterobacter	–	
	Bacillus	–	
Anaerobic	Clostridium	–	
Photosynthetic bacteria	Rhodospirillum	–	In aqueous systems
Symbiotic bacteria			
Associtated with leguminous plants	Rhizobium	Beans, peas, cloves and others	Nodules in roots or stems
Associtated with non Leguminous plants	Frankia	Alder, robinia, and others	Nodules in roots
	Azotobacter paspali		Colonies in roots or in
	Azoarcus sp.	Tropical or	soil nearby roots
	Spirillum lipoferum	subtropical grasses	
	Beijerinkia sp.		
Photosynthetic cyanobacteria			
Free-living	Anabaena		Soil surfaces,
	Nostoc sp.		Ponds, lakes
			In leaves or sprouts.
			Together with fungi in
Symbiotic	Anabaena	Azolla	lichens
	Nostoc sp.	Gunnera	

Biochemistry of N_2-fixation

The enzyme complex catalyzing the N_2-fixation of *Clostridium pasteurianum* was characterized to transfer six equivalent hydrogens to molecular N_2 and to release two molecules of H_2. The complex is extremely sensitive to oxygen, but as yet only the endproducts NH_3 and H_2 could be isolated. Intermediate products such as diimid (HN=NH) or hydrazine (H_2N-N_2H) are as yet only suspected.

Nitrogen-fixing bacteria fix N_2 at 20° C and 1 bar and the fixation is catalyzed by nitrogenase.

$$N_2 + 8\ H^+ + 8\ e^- + 16\ MgATP \rightarrow 2\ NH_3 + H_2 + 16\ MgADP + 16\ PO_4^{3-}$$

Nitrogenase is not specific for N_2, but will also reduce nitrous oxide (N_2O), cyanide (CN^-) and acetylene (C_2H_2) the reduction of acetylene to ethylene is particulary important as a tool to measure nitrogenase activity in a variety of ecosystems in the field or in the laboratory (acetylene reduction technique). Per mole of N_2 theoretically three moles of C_2H_2 are reduced:

$$N_2 + 6\,H \rightarrow 2\,NH_3$$
$$C_2H_2 + 2\,H \rightarrow C_2H_4$$

But this theoretical 3:1 ratio can vary considerably and sometimes amounts to 7:1 because different amounts of H_2 are released during N_2 fixation, but not during C_2H_2 reduction. The estimation of respiratory cost associated with the N_2 reduction in near natural conditions was achieved using simultaneously $^{14}CO_2$ and ^{15}N-labelling. Results indicate a minimum of 2.5 mg C per mg N fixed. This value was corrected by the estimation of the amount of carbon saved through the process of CO_2 fixation by the phosphoenolpyrophosphate (PEP) carboxylase of the nodules, using $^{14}CO_2$ in the soil atmosphere. This gives a real respiratory cost of 4 mg C per mg N fixed (Warembourg and Roumet, 1989).

The reduction of N_2 to NH_3 seems to be similar for all N_2-binding organisms, and the reductive enzyme always consists of two units of sulfur-iron proteins (ferredoxins) with molybdenum ions instead of iron ions in the second complex. Only both units together are able to reduce N_2 to NH_3 (Rees et al., 2005).

Establishment of the legume-rhizobium symbiosis involves Rhizobia´s attraction by root surfaces, where they attach to root cells by complementary macromolecules called *"lectins"* which are characterized as acylhomoserin lactones (Gonzales and Maketon, 2003). This phenomenon is called "quorum" sensing and is defined as the cell density-dependent regulation of gene expression. These autoinducers belong to a group of acylated homoserine lactones as signal molecules. They play a major role in preparing and coordinating the symbiotic nitrogen-fixing rhizobia during the establishment of their interactions with the host plant (Fig. 4.8, 4.9).

The initial response generally involves a tight curling of the root hair, and an infection thread is formed by enzymatic dissolution of the root hair cell wall to encase the invading rhizobia. The release of bacteria into cortical cells is followed by their rapid proliferation within the host cells (Fig. 4.10).

The plant root systems offer various microhabitats for bacterial growth and invasion by gram-negative diazotrophs, including *Azoarcus* in Kallar grass (a pioneer plant of saline soils in Pakistan and India) and in rice and *Acetobacter diazotrophicus* in sugar cane (Table 4.11). They all have the ability to grow deeply into plant roots, once inside the plant, they can spread systematically and reach aerial tissues via xylem vessels (Reinhold-Hurek and Hurek, 1998). It has been shown that some gramineous crops, such as Brazilean sugar cane or Kallar grass (in Pakistan and India) can grow without additional doses of nitrogen fertilizer.

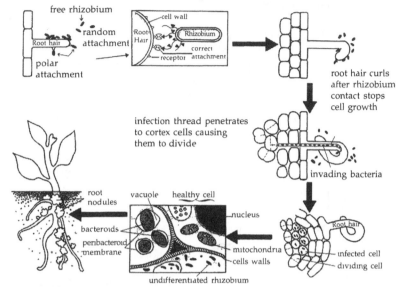

Fig. 4.8: Stages in the infection of legume roots by rhizobia (Ahmadijan and Paracer, 1986; see also Paul and Clarke, 1996).

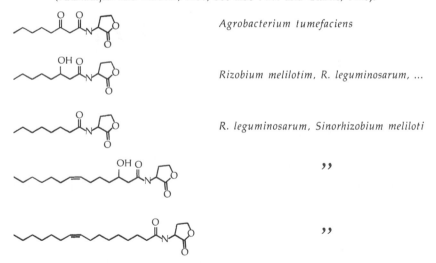

Fig. 4.9: Structures of acylhomoserine lactone to transfer information between root hairs and rhizobia (Gonzales and Maketon, 2003).

Increasing ammonium concentration decreases the activity of NH_3 formation and a 0.5 mM NH_4^+ concentration completely inhibits nitrogenase of *Azoarcus* sp. BH72 (Reinhold-Hurek and Hurek, 2000). Also two percent O_2 also diminishes fixation which is completely inhibited at 4% O_2 (Egener et al., 1999).

Table 4.11: Colonization of grass plants by diazotrophic endophytes (Reinhold-Hurek and Hurek 1998); see also Fig. 4.1

Microorganism	Plant	Location of bacterial infections
Azoarcus sp.	Rice, kallar grass	Roots, inter- and intracellular xylem vessels, shoots
Acetobacter diazotrophicus	Sugar cane	Roots, xylem vessels in lower stem
Herbaspirillum seropedicae	Sugar cane, sorghum	Roots and vessels
Herbaspirillum rubriscu albicans	Sugar cane, sorghum	Shoots and leaves, intracellular in xylem vessels

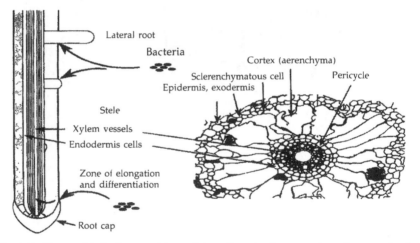

Fig. 4.10: Places of infections and bacterial settlings in the aerenchym from roots of grasses (Kallar grass and rice) (Reinhold-Hurek and Hurek, 1998).

Gaseous losses of nitrogen and controlling factors

Denitrification is the major process through which plant-available soil N is returned to various N-oxides and N_2:

+5		+3		+2		+1		0
Nitrate	$\rightarrow$	Nitrite	$\rightarrow$	Nitric	$\rightarrow$	½ Nitrous	$\rightarrow$	½
(NO_3^-)	- H_2O	(NO_2^-)	- ½ H_2O	oxide	- ½ H_2O	oxide	-H_2O	nitrogen
				(NO)		(N_2O)		(N_2)

This reaction can be simplified to:

$$2\,NO_3^- + 2\,H^- + 10\,H \rightarrow N_2 + 6\,H_2O$$

During denitrification by a bacterial respiratory process, nitrate or nitrite is reduced to nitrogenous oxides. All the redox reactions of denitrification are catalyzed by metalloenzymes, two of which contain copper. A review by Dooley and Chan (2006) described the structures, mechanisms, and assembly of these copper-containing enzymes: copper nitrite reductase and nitrous oxide reductase.

Reduction of nitrate to N_2O occurs at neutral pH at a redox potential of 500 to 600 mV, which is not essentially lower than the normal redox potential in soils of 700 to 800 mV. Reduction of N_2O to N_2 needs much lower redox potentials and starts only at potentials below 250 mV (Fig. 4.11).

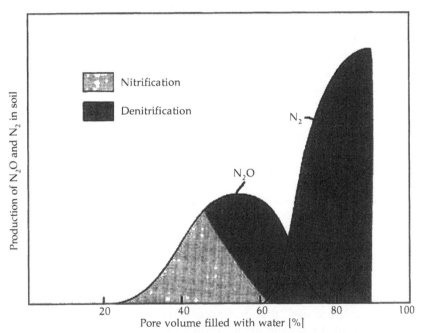

Fig. 4.11: Relation between water content in soils and the flow of N_2O and N_2 by nitrification and denitrification (Granli and Böckmann, 1994).

Denitrifying bacteria are diverse and include 13 genera of facultative anaerobes, but also several yeasts and filamentous fungi as well as nitrifying bacteria (Poth and Focht, 1985) (Table 4.12).

Girsch and de Vries (1997) isolated and characterized a NO reductase from *Paracoccus denitrificans*.

Thiobacillus denitrificans is a sulfur oxidizing bacterium that also reduces nitrate at anaerobic conditions, leading to an enrichment of sulfate in groundwater: It can substitute NO_3^- for O_2 as an electron acceptor:

$$2 \, NO_3^- + S + H_2O + Ca \, CO_3 \rightarrow Ca \, SO_4 + N_2$$

Table 4.12: Examples of denitrifying bacteria
(extracted from Haider, 1996).

Species	Ocurrence
Alcaligenes spp.	frequently occurring in soil
Agrobacterium spp.	some are also pathogenic for plants
Azospirillum spp.	N_2-fixing bacteria (also in grasses)
Bacillus spp.	some (also thermophilic)
Flavobacterium spp.	several denitrifying species
Propionibacterium spp.	denitrifying fermentative bacteria
Rhizobium spp.	N_2-fixing, but also denitrifying organisms
Thiobacillus spp.	chemoautotrophic S-oxidizing organisms

The N_2 is lost as a gas to the atmosphere.

The nitrous oxide (N_2O) reductase can be completely inhibited by acetylene (C_2H_2) and sulfide (HS^-). This inhibition (N_2O) provides a convenient method in studies of denitrification and quantifying N losses through denitrification (Benckiser et al. 1987).

Bacillus azotoformans can be cultivated under anaerobic conditions and is able to reduce completely nitrate to dinitrogen. From membranes of *B. azotoformans* a nitrous oxide reductase has been purified that consists of two subunits to transfer electrons to a dinuclear iron center where reduction of NO occurs (Suharti et al., 2001).

Nitrous oxide as an intermediate gaseous product of denitrification (and nitrification)

Formation of N_2O is explained by addition of NO to NO_2^-. This is discussed by Knowles (1981), resulting in doubling of N:

$$NO_2^- \quad \xrightarrow{NO} \quad N_2O \quad \rightarrow \quad N_2$$

N_2O is a "trace gas" that originates predominantly from biological sources. It results from both nitrification and denitrification (Fig. 4.12).

At redox potentials below -250 mV N_2O gets reduced to N_2:

$$N_2O + 2H^- + 2e^- \rightarrow N_2 + H_2O$$

Biochemistry of denitrification

Bacteria indicated in Table 4.6 need for reduction easily available C sources such as glucose or other sugars, fatty acids or proteins, several can even use methane or sulfur compounds. Chemolithotrophs such as *Alcaligenes eutrophus* use H_2 or reduced sulfur compounds. Several *Pseudomonas* spp. are able to degrade under anaerobic conditions chlorinated or nitrated aromatic compounds.

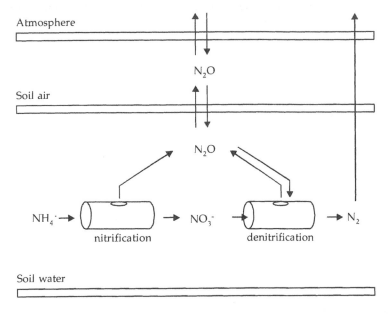

Fig. 4.12: N_2O production during nitrification and denitrification (Davidson and Kinglee, 1997).

The enzyme nitrous oxide reductase catalyzes the two-electron reduction of N_2O to N_2 (Fig. 4.12). The nitrous oxide reductase from *Paracoccus denitrificans* and *Pseudomonas* spp is a dimer and contains two copper centers. The N-terminal domain contains the center to catalyze the reduction in the form of a copper-sulfide cluster.

Due to its long mean residence time in the atmosphere (150 years) N_2O is a trace gas with a greenhouse potential 206 times greater than that of CO_2 and with that great potential is participating in the present climatic change (Duxbury, 1994). Although it contributes today only 5% of this potential, considering the tremendous increase in fertilizer consumption it might in the future even surpass the present activity of CO_2 and CH_4 (Table 4.13, Fig. 4.13).

N_2O is a very stable trace gas that is removed only by photolysis in the stratosphere. Furthermore, it has a capacity to decompose ozone to activated oxygen radicals (Crutzen and Ehalt, 1997).

$$N_2O \quad \xrightarrow{\nu < 240 \text{ nm}} \quad N_2 + O^x$$

Table 4.13: Global atmospheric emission of N_2O [Tg N y^{-1}] (Galloway et al., 2004).

	1860	Early 1990s	2050
Soils			
Natural	6.6	6.6	6.6
Anthropogenic	1.4	3.2	3.2 ± ?
Rivers			
Natural	0.05	0.05	0.05
Anthropogenic	–	1.05	3.32
Estuaries			
Natural	0.02	0.02	0.02
Antropogenic	–	0.2	0.9
Shelves			
Natural	0.4	0.4	0.4
Antropogenic	–	0.2	0.32
Ocean (natural)	3.5	3.5	3.5
Total	12	15.2	18.2 ± ?

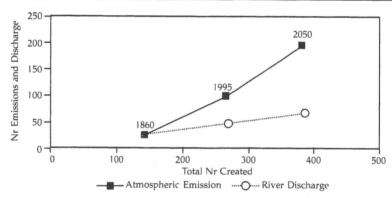

Fig. 4.13: Atmospheric emission of reactive nitrogen (Nr), such as NOx and NH_3, and river discharge as a function of N-fertilization in Tg N y^{-1} (Galloway et al., 2004).

Global NH_3 losses from fertilizers and manures

One of the reasons for the low efficiency in nitrogen use by crops is the volatilization of ammonia from synthetic fertilizers and from animal manures. The calculated median NH_3 losses from fertilizer (78 Mt yr^{-1}) and animal manure N (33 Mt yr^{-1}) amount to 14% of the nitrogen contained in fertilizers and 23% of N in animal manure. The estimated NH_3 loss from animal manure is 21% in industrialized and 26% in developing countries. NH_3 is predominantly released from domestic animal manure and urea fertilization (Schlesinger and Hartley, 1992; Table 4.7).

According to Bouwman et al. (2002), Schlesinger and Hartley (1992) and Dentener et al. (2006) the major sources of global ammonia volatilization loss include excreta of domestic animals at 21.6 Tg N yr^{-1} and wild animals at 0.1 Tg N yr^{-1}, use of synthetic fertilizers at 9 Tg yr^{-1}, oceans at 8.2 Tg yr^{-1}, biomass burning 5.9 Tg N yr^{-1}, crops at 3.6 Tg N yr^{-1}, human population and pets at 2.6 Tg yr^{-1}, soils under natural vegetation 2.4 Tg yr^{-1}, industrial processes 0.2 Tg yr^{-1} and fossil fuels 0.1 Tg N yr^{-1} (Table 4.7). About half of the global emissions come from Asia and about 70% from food production and are located in the Indian subcontinent and China reflecting the density of animals and the intensity of fertilizer use.

NO$_x$ and NH$_3$ are deposited as dry or wet precipitation back on vegetation, soils, and aqueous resources, but have no climatic effect as greenhouse gases.

The application of synthetic fertilizer is known to result in the release of NH$_3$ into the atmosphere. The degree of realease is a function of the type of fertilizer, soil properties, temperature, wind speed, precipitation and agricultural management. In the developing countries about 55% of the fertilizer used is in the form urea, because of its relatively high N content resulting in low transportation costs. The emission of NH$_3$ in tropical regions is about 2.5 times that of temperate regions because higher temperatures result in larger losses and because of the widespread use of urea and NH$_4$HCO$_3$ (Tab. 4.14).

Emission of ammonia corresponds to amounts of 40 Mt N yr^{-1}. This corresponds to about half of the commercial fertilizer use (see Table 4.7). Agriculture and especially liquid manure contribute more than 95% to the release of ammonia. Measurement in a forest area in the Netherlands indicated NH$_4^+$ depositions of 17 kg N ha^{-1} yr^{-1} by rain and 42 kg N ha^{-1} yr^{-1} by release from leaf surfaces of trees. This additional N-deposit causes disturbance of the equilibrium between N deposition and N uptake of plants.

Turnover of Phosphorus in Soil

Phosphorus is also a macronutrient for plants and a primary component of biomolecules and has an essential function in molecules such as adenosine triphosphate, deoxyribonucleic acid, phospholipids, the phosphopyridine nucleotides, and others. Plant requirements for P are on average between 0.2 and 0.4% of plant dry weight.

The original sources of most soil P are the mineral apatite or rock phosphate. Both of these forms are highly insoluble. The soluble and plant-available form of P in soils is orthophosphate, H$_2$PO$_4^-$ or HPO$_4^{2-}$, which forms precipitates as Al-, Fe-, or Ca-phosphates or is sorbed by Al- and Fe oxides. Therefore, soluble P is present at very low concentrations in soil solutions and must be continuously replenished (Stewart and McKercher, 1982); see Fig. 4.14.

The microbial biomass participates actively in solubilization of relatively insoluble forms of Al , Fe- and Ca phosphates. It has been

Table 4.14: *Top*: Consumption of synthetic fertilizers and ammonia emissions for temperate and tropical zones (based on estimates from Internat. Fertilizer Ind. Assoc. (1994) and Bouwman et al. (1997)). *Bottom*: Sources and sinks of atmospheric NH_3 (according to Schlesinger and Hartley, 1992).

Fertilizer type	Global consumption $[Tg^* \; N \; yr^{-1}]$	Temperate zones $[Gg^* \; N \; yr^{-1}]$	Tropics $[Gg \; N \; yr^{-1}]$	Total $[Gg \; N \; yr^{-1}]$
Ammonium sulfate	2.6	34	169	203
Urea	29.2	1,632	4,137	5,769
Ammonium nitrate	8.2	25	141	166
Calcium ammonium nitrate	4.1	9	72	82
Anhydrous ammonia	52	18	190	208
Nitrogen solutions	4.2	11	93	104
Ammonium phosphates	3.7	35	113	147
Other NP-N	3.2	18	77	95
No N-type	0.4	2	7	9
Total	77.0	2,626	6,409	9,035

*(1 Gg = 10^9 gram; 1 Tg = 10^{12} gram)

Sources	Mean$[Tg \; N \; y^{-1}]$	Range$[Tg \; N \; y^{-1}]$
Domestic animals	32	24 to 40
Sea surface	13	8 to 18
Forests, Savannah, Prairie soils	10	6 to 45
Fertilized soils	9	5 to 10
Biomass combustion	5	1 to 9
Human excretions	4	
Combustion of coal	2	
Traffic	0.2	
Σ of Sources	75	50 to 128

Sinks		
Deposits on dry and wet surfaces	30	10 to 30
Sea surface	16	
Reaction with OH-radicals	1	
Σ of Sinks	57	

demonstrated that various kinds of enzymes, e.g., the phosphatases, are released also by plant roots and microorganisms. The most common enzymes in soils are polyphenol oxidases and phosphatases. The latter allow the production of inorganic phosphate, the only form available for plant roots and soil microorganisms. Phosphate solubilization has been observed for many different species of bacteria, fungi, and actinomycetes. The release of soluble inorganic P from organically bound P indicated that microbial P solubilization should be a possible way to increase P availability for plants (Illmer and Schinner, 1995).

Organic P can account for 3 to 90% of the total soil P, with 30 to 50% common in most soils. Total organic P is a function of organic matter content, and organic matter increase is parallelled by organic P increase.

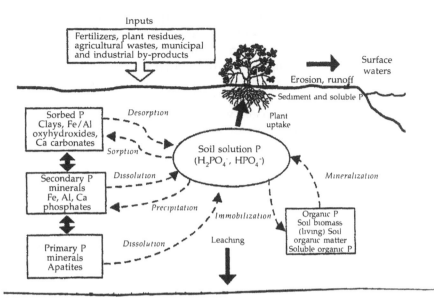

Fig. 4.14: P fractions and fluxes in soil (Stewart and McKercher, 1982).

Role of mineral colloids in cycling of P and other minerals

To compare concentrations of SOM in different size separates, Guggenberger and Haider (2002) defined the enrichment factor E, which relates the content of an organic matter or mineral content in a particular size fraction to that in the whole soil, e.g.,

$$E_{C,N,P,S} = (mg\ C,\ N,\ P,\ or\ S\ g^{-1}\ fraction)/mg\ C,\ N,\ P,\ or\ S\ g^{-1}\ whole\ soil)$$

Examples for the enrichment factors of organic C (EOC), total N (EN), and organic P (EOP) are shown in Table 4.15.

Table 4.15: Enrichment factors (E) for organic C, total N, and organic P in particle-size separates of topsoils (adapted from Guggenberger and Haider, 2002).

USDA taxonomy	Site	Particle size [μm]	EOC	EN	EOP
Alfisol	Arable field(Denmark)	> 2	5.04	6.18	5.38
Inceptisol	Arable field Bavaria, Germany	> 2	2.51	2.39	2.06
Inceptisol	Arable field Bavaria, Germany	2 to 20	1.00	0.83	0.66
		20 to 2000	0.26	0.16	0.13

Generally the highest enrichment factor for C is found in fine silt- and clay-size separates and is similar for total N and organic P. The major form of phosphorus in soil is bound to organic structes. The distribution of different organic forms of P in different particle size separates has also been examined by means of ^{31}P-NMR spectroscopy of alkaline extracts.

Fig. 4.15 shows that compared to bulk soil, diester-P structures are enriched in the clay fraction, whereas monoester P structures are enriched in particle size classes coarser than clay. Because the major organic P structure in microorganisms is diester-P, including teichoic acid P, this indicates that organic P associated with clay is also primarily of microbial origin. There is evidence that diester P produced by microorganisms can be stabilized on mineral surfaces with Fe oxides being most effective (Germida and Siciliano, 2000).

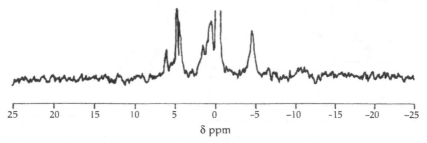

δ ppm

Chemical shifts and structural assignments:

ppm	
–3,8 to –5,5	pyrophosphate
–0,4 to 1, 0	P-diester (phospholipids and DNA)
3,0 to 1, 0	Teichoic acid
5,7 to 3,0	P-monoester (inositolphosphate, sugarphosphate, mononucleotides, ...)
6,7 to 5,7	inorganic PO_4^{3-} (orthophosphate)

Fig. 4.15: ^{31}P-NMR spectrum of extracted birch litter compost.

Turnover of sulfur in soil

Similar to phosphorus, sulfur is an essential element for the growth and activity of organisms. It exists in a number of oxidation states (+6 to –2), and the most oxidized and most reduced states are the important components required for synthesis of essential compounds such as amino acids, proteins, enzymes, and coenzymes active in energy transfer. Sulfur plays an important role in many enzymatic processes and even in the evolution of life (Kroneck, 2005). Representative examples are the copper enzymes dinitrogenmonooxide reductase for reduction of N_2O, the iron molybdenum enzyme nitrogenase in binding N_2, and the cytochrome nitrite reductase in reducing nitrite.

Sulfur is considered as a macronutrient in most ecosystems, but sometimes its availability is limiting: less than 100×10^6 ha of sulfur deficient soils are found in many parts of the world. The S-concentration in the surface layers varies for European- and US-soils from 27 to 1100 μg per g soil (Germida et al., 2002).

Inorganic sulfur accounts in most soils to less than 25% (Table 4.9) and constitutes more than 90% of the total S present. It can be grouped broadly in organic sulfates and C-sulfide-structures, e.g. in the amino acids cysteine, cystine, and methionine, respectively. Organic sulfates include sulfate esters (C-O-S), sulfamates (C-N-S), and sulfated thioglycosides (N-O-S). Mineralization of such S-compounds is controlled by either biological or biochemical pathways. Biological mineralization is controlled by the microbial need for C and energy, whereas biochemical mineralization is influenced by enzyme synthesis, activity and kinetics.

Microbial reactions clearly dominate the oxidative conversion; microorganisms involved can be divided into (1) chemolithotrophs (*Thiobacillus* sp.), (2) photoautotrophs including species of purple and green sulfur bacteria, and (3) heterotrophs, including a wide range of bacteria and fungi (Wainwright, 1984). Plants significantly increase S-mineralization in soils because of the increased availability of energy sources in the rhizosphere (Speir et al., 1980).

Microbial and biochemical S transformation reactions are demonstrated in Fig. 4.16.

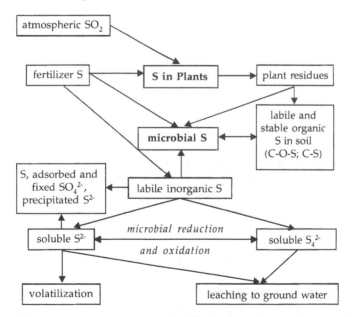

Fig. 4.16: Main transformation reactions of sulfur in the soil-plant-microorganism system (Germida et al., 2002).

Reduction of oxidized inorganic sulfur compounds

The incorporation of sulfur into cellular compounds is caused by assimilatory sulfur reduction which can be conducted by many organisms. It commonly is the first step in sulfate activation, but not connected to the respiratory chain (Schlegel, 1992).

The dissimilatory reduction of sulfur (sulfate respiration) is used by numerous obligate anaerobic and organotrophic organisms that use SO_4^{2-} (similar to nitrate during nitrate respiration or denitrification) as an electron acceptor.

Oxidation of reduced inorganic sulfur compounds

Chemolithotrophic bacteria gaining energy by oxidation of reduced S compounds belong to the *Thiobacillus* genera. They grow facultatively under anaerobic conditions, fix CO_2, and can also use organic compounds and nitrate as electron acceptors (*Thiobacillus denitrificans*). Others (*Thiobacillus ferrooxidans*) also use besides reduced S compounds Fe^{2+} as an energy source in soils or sediments by oxidizing Fe^{2+} to Fe^{3+}. Oxidation of sulfur compounds to sulfuric acid might lead to acidification of soils or water bodies, which is used in mining regions to solubilize metals from ores (metal leaching).

Phototrophic sulfur purple bacteria or green sulfur bacteria (genera *Chromatium* and *Chlorobium*) can oxidize H_2S in deeper water layers or sediments to sulfur and further to sulfate. They use reduction equivalents from light energy to assimilate CO_2.

$$CO_2 + H_2S \xrightarrow{\text{Light}} |CH_2O| + 2\,S + H_2O$$

$$3\,CO_2 + 2\,S + 5\,H_2O \xrightarrow{\text{Light}} 3\,|CH_2O| + 2\,H_2SO_4$$

These phototrophic bacteria occur in rice fields or anaerobic ponds and are scarcely important for the oxidation of sulfur in aerobic soils.

Heterotrophic bacteria belonging to the genera *Arthrobacter*, *Bacillus*, *Micrococcus*, *Mycobacterium*, or *Pseudomonas* and fungi can oxidize sulfur in aerobic soils but cannot gain energy and use organic compounds as substrates (Germida et al., 2002).

Sulfur is an essential element for the growth and activity of organisms. It exists in a number of oxidation states, i.e., +6 to –2, the most oxidized and reduced state that can be taken up by organisms. Plants need substantial amounts of S for growth and grain production. It is important to understand the nature and quantities of different S pools in soil and the various transformation processes of the S cycle. Many of the transformation of S compounds are mediated by microbial activity (Fig. 4.16).

REFERENCES

Ahmadijan V, Paracer S, 1986: Symbioses: An introduction to biological systems. University Press of New England, Hannover.

Almendros G, Fründ R, Gonzalez-Vila FJ, Haider K, Knicker H, Lüdemann HD, 1995: Analysis of ^{13}C- and ^{15}N CPMAS-spectra of soil organic matter and composts. FEBS Letters 282, 119–121.

Anderson HA, Bick W, Hepburn A, Stewart M, 1989: Nitrogen in humic substances: In: Humic Substances II: In Search of Structure, M.B.H. Hayes et al. eds., pp. 223–253, J. Wiley and Sons, Chichester.

Benckiser G, Gaus G, Syring KM, Haider K, Sauerbeck D, 1987: Denitrification losses from an Inceptisol field treated with mineral fertilizer or sewage sludge, Z Pflanzenernähr. Bodenkd. 159: 421–428.

Bouwman AF, Bouman LJM, Batjes NH, 2002: Estimation of global NH_3 volatilization loss from synthetic fertilizers and animal manure applied to arable land and grasslands. Global Biogeochem. Cycles 16, Art. No. 1024.

Bouwman AF, Lee DS, Asman WAH, Dentener FJ, van der Hoek KW, Olivier JGJ, A global high-resolution emission inventory for ammonia. Global Biogeochem. Cycles 11, 561–578, 2002.

Burke IC, Yonker CM, Parton WJ, Cole CV, Flach K, Schimel DS, 1989: Texture, climatic and cultivation effects on soil organic matter content in US grassland soils. Soil Sci. Soc. Amer. J. 53, 800–805.

Burns RC, Hardy RWF, 1975: Nitrogen fixation in bacteria and higher plants. Springer, Berlin, NewYork.

Cleveland CC, Townsend AR, Schimel DS , Fisher H, Howart RW, Hedin LO, Perakis SS, Latty EF, Fisher JC, Elseroad A, Wasson MF, 1999: Global patterns of terrestrial biological nitrogen (N_2) fixation in ecosystems. Global Biogeochem. Cycles 13, 623–645.

Cowling EB, 2002: Reactive nitrogen and the world: 200 years of change. Ambio. 31, 64–71.

Crutzen PJ, Ehalt DW, 1997: Effects of nitrogen fertilizer and combustion on the stratospheric ozone layer. Ambio. 6, 117–122.

Davidson EA, Kinglee W, 1997: A global inventory of nitric oxide emission from soils. Nutr. Cycl. Agroecosyst. 48, 37–50.

Dentener F, Stevenson D, Ellingsen K, van Noije T, Schultz M, Amann M, Atherton C, Bell N, Bergmann D, Bey I, Bouwman L, Butler T, Cofala J, Collins B, Drevet J, Doherty R, Eickhout B, Eskes H, Fiore A, Gauss M, Hauglustaine D, Horowitz L, Isaksen ISA, Josse B, Lawrence M, Krol M, Lamarque JF, Montanaro V, Muller JF, Peuch VH, Pitari G, Pyle J, Rast S, Rodriguez J, Sanderson M, Savage NH, Shindell D, Strahan S, Szopa S, Sudo K, Van Dingenen R, Wild O, Zeng G, 2006: The global atmospheric environment for the next generation. Environ. Sci. Technol. 40, 3586–3594.

Dobermann A, Cassman KG, 2004: Environmental dimensions of fertilizer Nitrogen: What can be done to increase nitrogen use efficiency and ensure global food securiy. Plant Soil 247, 25–39.

Dooley DM, Chan J M, 2006: Copper enzymes in denitrification: Enzycloped. Inorgan. Chem., Wiley, Chichester.

Duxbury JM, 1994: The significance of agricultural sources of green house gases. Fertilizer Res. 38, 151–163.

Egener T, Hurek T, Reinhold-Hurek B, 1999: Endophytic expression of nif genes of *Azoarcus* sp strains BH72 in rice roots. Mol. Plant-Microbiol Interact. 12, 813–819.

Evans LT, 1998: Feeding the 10 billions, Cambridge Univ. Press, 128 pp.

Fründ R, Lüdemann HD, 1989: The quantitative analysis of solution and CPMAS-C-13-NMR spectra of humic material. Sci. Total Environ. 81/82, 157–168.

74   Soil Biochemistry

Galloway JN, Dentener FJ, Capone DG, Boyer EW, Howarth RW, Seitzinger GP, Asner
GP, Clrveland CC, Green PA, Holland EA, Karl DM, Michaels AF, Karl DM, Michaels
AF, Porter, JH, Townsend AR, Vörösmarty GJ, 2004: Nitrogen cycles: Past, present,
and future. Biogeochem. 70, 153–226.

Germida JJ, Wainwright M, Gupta VVSR, 2002: Biochemistry of sulfur cycling in soil. Soil
Biochem. 7, 1–53.

Germida JJ and Siciliano SD, 2000: Phosphorus, sulfur and metal transformations. In: M.
Sumner (ed.). Handbook of Soil Sci. pp. 95 –106, Boca Raton Fl, CRC Press.

Girsch P, de Vries S, 1997: Purification and initial kinetic and spectroscopy characterization
of NO reductase Paracoccus denitrificans. Biochim. Biophys. Acta 3318: 202–216.

Granli T, Böckman OC, 1994: Nitrous oxide from agriculture. Norweg. J. Agric.Sci. Suppl.
12, 1–127.

Green PA, Holland EA, Karl DM, Michaels AF, Porter JH, Townsend AR, Vörösmarty CJ,
2004: Nitrogen cycles: past, present, and futures. Biogeochem. 70, 153–226.

Gonzales JE, Marketon MM, 2003: Quorum sensing in nitrogen fixing Rhizobia. Microbiol.
and Molec. Biol. Rev. 67, 574–592.

Guggenberger G, Haider K, 2002: Effect of mineral colloids on biogeochemical cycling of
C, N, P and S in soil. In: Interactions between soil particles and microorganisms.
IUPAC Series on Analytical and Physical Chemistry, Vol. 8, Ch. 7, 267–322. John
Wiley and Sons. Ltd., New York.

Haider K, 1996: Biochemie des Bodens. Enke Verlag, Stuttgart, 174 pp.

IFA/FAO (International Fertilizer Industry Association/Food and Agricultural
Organization), 2001: Global estimates of gaseous emissions of NH_3 , NO and N_2O
from agricultural land.

Illmer P, Schinner F, 1995: Solubilization of inorganic calcium phosphartes—solubilization
mechanisms. Soil Biol. Biochem. 27, 257–263.

Knicker H, Lüdemann H-D, Haider K, 1997: Incorporation studies of NH_4^+ during
incubation of organic residues by ^{15}N-CPMAS-NMR spectroscopy. Europ J. Soil Sci.
48, 431–441.

Knowles R, 1981: Denitrification, In Soil Biochem. Vol. 5, E.A. Paul, J. Ladd (eds.). M Dekker,
New York, p 323 –369.

Kroeze C, Mosier A, Bowman L, 1999: Closing the global N20 budget: a retrospective
analysis 1500–1994. Global Biogeochem. Cycles 13, 1–8.

Kroneck PMH, 2005: The Biochemical Cycles and the evolution of Life: In: Sigel et al. Eds.:
Metal ions in Biological Systems. Marcel Dekker, Basel, 43, 1–7.

Malcolm RL, 1989: Application of solid state 13C NMR spectroscopy to geochemical studies
of humic substances. In: Humic Substances II, M.B.H. Hayes et al. eds. Wiley and
Sons, Chichester, p. 339–372.

Miltner A, Haumaier L, Zech W, 1998: Transformation of phosphorus during incubation
of beech leaf litter in the presence of oxides. Eur. J. Soil Sci. 49, 471–475.

Mohr H, 1994: N deposition causing new forest damages. Spektrum Wissensch., Jan. 1994,
48–53.

Mosier A, Kroeze C, 2000: Potential impact on the global atmospheric N_2O budget of the
increased nitrogen input required to meet future global food demands. Chemophere—
Global Change Sci. 2, 465–473.

Olivares FL, 1998: Infection and colonization of sugar cane and other gramineous plants
by endophytic diazotrophs. Crit. Rev. Plant Sci. 17, 77–119.

Paul EA, Clark FE, 1996: Soil Microbiology and Biochemistry. Second Edition. Academic
Press, San Diego, Cal., 340 p.

Peoples MB, Boyer EB, Goulding KWT, Heffer P, Ochwoh VA, Vanlauwe B, Wood S, Yagi
K van, Cleemput O, 2004: Pathways of nitrogen Loss and their impacts and human
health and the environment. In: Agriculture and the Nitrogen Cycle, Scope 65, A.R.
Mosier, J.K. Syers and J. Freney eds., Island Press, 53–69.

Poth M, Focht DD, 1985: ^{15}N kinetic analysis of N_2O production by *Nitrosomonas europaea*: an examination of nitrifyer denitrification. Appl. Environ. Microbiol. 49,1134–1141.

Rochester IJ, Peoples MB, Hulugalle NR, Gault RR, Constable GA, 2001: Using legumes to enhance nitrogen fertility and improve soil condition in cotton cropping systems. Filed Crops Res. 70, 27–41.

Quispel A, 1974: The biology of nitrogen fixation. North Holland, Amsterdam, 769 pp.

Rees DC, Teczan FA, Haynes CA, Walton MY, Andrade SL, Einsle O, Howard JB, 2005: Structural basis of nitrogen fixation. Phil. Trans. R. Soc. A 363, 971–984.

Reinhold-Hurek B, Hurek T, 1998; Life in grasses: diazotrophic endophytes. Trends Microbiol. 6:139–144.

Reinhold-Hurek B, Hurek T, 2000: Reassessment of the taxonomic structure of the diazotrophic genus *Azoarcus sensu lato* and description of three new genera and species, *Azovibrio restrictus* gen. nov. sp. Nov., and *Azospira oryzae* gen. nov. sp. Nov. and *Azonexus funguphilus* gen. nov. sp. Int. J. Syst. Evol. Microbiol. 60, 649–659.

Roberts TL, Bettany JR, Stewart JWB, 1989: A hierarchical approach to the study of organic C, N; P, S in western Canadian soils. Can. J. Soil Sci. 69, 739–749.

Schnitzer M, 1985: Nature of nitrogen of humic substances. in: Humic Substances in Soil, Sediments, and Water. G.R. Aiken, D.M. Mc Kknight, R.L. Wershaw P.T., eds. Wiley and Sons. Chichester.

Schlesinger WH, 1991: The global carbon cycle. In: Schlesinger WH. Biogeochemistry: an analysis of global change. Acad Press, San Diego, p. 308–320.

Schlesinger WH, Hartley AE, 1992: A global budget for atmospheric ammonia. Biogeochem. 15, 191–211.

Schulten HR, Sorge C, Schnitzer M, 1995: Structural studies on soil nitrogen by Curie-point pyrolysis-gas chromatography/mass spectrometry with nitrogen selective detection. Biol Fertil. Soils 20, 174–184.

Seitzinger SP, Styles RV, Boyer E, Alexander RB, Billen G, Howarth R, Mayer B, Van Breemen N, 2002: Nitrogen retention in rivers: model development and application to water sheds in the eastern US. Biogeochem. 57/58, 199–237.

Speir TW, Lee R, Pansier EA, Cairns A, 1980: A comparison of sulfatase, urease, and protease activities in planted and in fallow soils. Soil Biol. Biochem. 12, 281–291.

Stewart JWB, McKercher RB, 1982: Phosphorus cycle. In: Experimental Microbial Ecol. Chapter 14: 221–238.

Suharti MJ, Strampraad I, Schroder S, 2001: A novel copper A containing menaquinol NO reductase from *Bacillus azotoformans*. Biochem. 40, 2632–2639.

Tate R, 1987: Soil organic matter, Wiley and Sons, New York, 291 pp.

Wainwright M, 1984: Sulfur oxidation in soils. Adv. Agron. 37, 349–396.

Warembourg FR, Roumet C, 1989: Why and how to estimate of symbiotic N_2 fixation? A progressive approach based on the use of ^{14}C and ^{15}N isotopes. In: Ecology of arable land, M. Clarholm and L. Bergström eds., Kluwer Academic Publishers, pp. 31–41.

Winogradsky S, 1890: Recherche sur les organismes de la nitrification. Ann. Inst. Pasteur 4, 257–275.

5

Composting and Fermentation of Organic Materials

The increasing price for crude oil makes the use of biomass and of municipal biowaste and biomass from agriculture and forestry for recovery of energy more favourable. They can be used as solid fuel for heating plants, as liquid fuel in automotives and engines or as biogas derived from manure or biowaste. It can be fed in gas grids. For heating use wood from forestry is the most important solid fuel for heating plants, especially in smaller communities. A similar rapid development can be ascertained for liquid fuels from rapeseed as well as grain, and maize as feedstock for both, bioethanol and biogas (Widmann, 2007; Quicker and Faulstich, 2007).

Nature of biowaste

Solid waste consists of items such as product packaging, garden clippings, food scraps, waste papers, rubbish, and household and agricultural refuse. On average about 80% consists of biodegradable organic materials. Materials of organic origin are known as biomass. Incineration or landfill are described as methods to reduce its portion, but biotechnology techniques also offer opportunities to lower the costs of other methods such as incineration, and are a way to better protect the environment by creating valuable products from biowaste, such as composts as fertilizers or biogas as fuels.

Lignocellulosic biomass represents a primary fraction of municipal solid waste (Table 5.1).

The efficiency of biotechnological methods for waste treatment is based on the capacity of organisms to degrade organic materials and to absorb hazardous substances (Haider, 1999; Fischbeck, 1993).

Table 5.1: Average composition of municipal solid waste (Wyk, 2001).

Constituent	Weight (%)
Paper and paper products	37.8
Food waste	14.2
Yard waste	14.6
Wood waste	3.0
Total: Cellulose waste	**69.6**
Plastic	4.6
Rubber and leather	2.2
Textiles	3.3
Glass and ceramics	9.0
Metals	8.2
Miscellaneous	3.1
Total	**100**

Aerobic waste disposal by composting

Chapter 2 described the principles of degradation and conversion of biopolymers such as cellulose, polysaccharides, and lignin of plants by soil microorganisms. Similarly, the biopolymers in biowastes can be managed and transferred by aerobic composting or fermentation into organic fertilizer or combustible biogas.

A common method of dealing aerobically with biowaste is composting. This use of byproducts of vegetable, animal and human origin has been known for over 2000 years since the Roman Empire and describes how organic waste has to be processed before it can be used in agricultural soils. However, since World War II the use of waste for land fertilization has decreased and farmers in developed countries have markedly increased the use of mineral fertilzer instead of organic amendments. Concurrently, the amounts of materials from municipal solid waste, sewage sludges and wastes of agroindustrial origins have increased exponentially, and millions of tons of organic matter are landfilled or incinerated.

The decomposition of matter by microorganisms (bacteria and fungi) produces a humuslike substrate that can be used as fertilizer, also acts as a soil conditioner because it has been stabilized, decomposes slowly, and thereby remains effective over a long time period.

Degradation of the raw material follows first-order kinetics. During the first few days and weeks the oxygen demand and heat production are greatest, when degradation of organic residues is at the maximum. If oxygen becomes limiting during the thermophilic phase organic acids are formed by anaerobic fermentation; these accumulate but are degraded as oxygen becomes available. If they persist, their phytotoxicity can affect the quality of the compost product.

During the thermophilic phase, compost can lose considerable quantities of nitrogen ranging from less than 10% to greater than 60% of the initial N content. Losses include volatization of NH_3, N_2 and N_2O. Total N losses are inversely correlated with the C/N ratio and are minimized for C/N greater than 30:1 (Rynk and Richard, 2001).

After a thermophilic phase ranging from a few weeks to 2 months, temperatures decline to the mesophilic range, when easily degradable compounds become increasibly scarce. The compost then begins to "cure".

Composts affect several soil physical functions, including water transport and storage, gas exchange, and heat transfer. The presence of organic matter in composts stimulates changes in microbial activity of the soils: because carbon is the dominant element in composts and the resulting biomolecules have additional ion exchange sites, addition of compost to soils will increase its overall cation exchange capacity and thus its ability to chelate minerals and heavy metals (Dick and Mc Coy, 1993).

The application of composts to soil affects a series of biological, chemical, and physical transformations as well as soil properties and processes. It affects water transport and storage, gas exchange and heat transfer, inhibits erosion, promotes soil aggregation, and enhances aggregate stability. Many of these improvements peak within a few weeks or months after application of composts and decline afterwards. Longer lasting effects on soil properties can be achieved by repeated compost applications over a period of years.

Microbiology of composting

Microbiologists divide the microbial population of compost into bacteria, actinomycetes, and fungi. Because high temperatures of 40 to 70 °C are characteristic mainly at the beginning of composting, the thermophilic representatives of each group have more attention than the mesophilic ones, which, however, are as important in the composting process (Kutzner, 2000).

Studies on thermophilic actinomycetes in self-heated organic material such as hay, manure, and composts go back to Waksman (1926, 1931) and his interests in the formation of humus.

Emission of microorganisms from composting plants and other undesirable side effects of compostings

A problem facing any composting method is changes in composition that make compost undesirable for land application. It is desirable that composts are free of pathogenic agents and do not impair plant growth. Fresh, immature composts have often been observed to exert toxic effects on plant growth. Therefore, a test on phytotoxicity using seeds of cress or radish is used routinely to prove compost maturity, as is a test for the presence of volatile acids (e.g. acetic, propionic, and butyric acids or other toxic components) in immature compost (Mathur et al., 1993; Baffi et al., 2007). The production of compost also leads to the emission of odors and they are often the cause of the objections by the neighborhood against newly installed compost plants (Haug 1993).

It is apparent that persons employed in composting plants or coming into contact with the material may be adversely affected by pathogenic agents contained in these materials or developed during the composting process. The emission of germs via dust has become a major concern of environmental microbiologists and public health officers. However, if care is taken to protect employees and the close neighborhood, there apparently exists no specific risk originating from compost plants (Epstein and Epstein, 1989).

Millner et al. (1994) concluded that composting poses no significant level of biohazard risk to the health and welfare of the general public. This conclusion was based on the fact that compost workers during a 10-year period at a composting site showed no adverse health impacts.

Phytopathogenic aspects of composting

Because of the sources of starting materials (dead plants, various household wastes, and others) and the application of the produced compost as growth substrate for plants, the composts sold for this purpose have to be checked for the absence of soilborne pathogens, and the composting process itself has to be verified to be effective in inactivating selective phytopathogenic test organisms e.g. *Plasmodiophora brassicae, Fusarium oxysporum, Phytium ultimum, Sclerotiuia sclerotiorum*, tobacco mosaic virus and others. The literature on the inactivation of phytopathogenic organisms by composting has been reviewed by Bollen (1993) and by Bollen and Volkers (1996).

Degradation of more recalcitrant compounds during composting

Degradation of organic matter during composting is conducted mostly by bacterial and fungal populations. Their ability to degrade stable compounds such as humic compounds or xenobiotics is limited. Compost, however, has an enormous potential for bioremediation because its diverse population of microorganisms is able to degrade a variety of compounds including contaminants such as polycyclic aromatic hydrocarbons, diesel fuel (Namkoong et al., 2002), and other compounds (see also Table 2.2). Polycyclic aromatic hydrocarbons occur in natural media such as soils and sediments (Scelza et al., 2007). Their main source is the incomplete combustion of coal and other fossil fuels.

In the global carbon cycle white rot fungi play a crucial role in the transformation and degradation of relatively recalcitrant organic compounds due to their ligninolytic enzymes. In particular manganese peroxidase, lignin peroxidase, and laccase play a crucial role in this respect (see chapter 2). Various actinomycetes are also able to utilize lignin monomers as carbon and energy source and to a limited extend the polymer itself.

The mechanism of lignin degradation by actinomycetes has been extensively studied at the enzyme level using *Streptomyces viridoporus*, and appears to differ from that by fungi. It produces water-soluble, acid-precipitable compounds as end products (Crawford, 1988).

The degradation of lignin appears to be a domain particular to basidiomycetes, especially *Phanerochaete chrysosporium*, a thermotolerant species that has been isolated from wood chip piles. It is one of the most active lignin degraders. Thus the microbial population of the late thermophilic and the following maturation phases consist of a high proportion of degraders of recalcitrant macromolecules, including lignin, whereas the early population degrades the more readily available nutrients.

Anaerobic fermentation of waste—production of methane and its use as an energy source

Today most biological wastes are composted, and this technology is already well developed, but anaerobic processes also gain increasing importance for the utilization of solid organic wastes and have been successfully applied as treatments of wastewater, sewage sludge, and manures. Anaerobic digestion of municipal solid waste is a relative young technique developed in the last 10 to 15 years (Wyk, 2001).

Anaerobic degradation is effected by specialized various groups of bacteria in several successive steps, and each step depends on the preceding one. The whole anaerobic fermentation process can be divided into three steps (Fig. 5.1):

During hydrolysis, the mostly water-insoluble biopolymers such as carbohydrates, proteins, fats, and others, are enzymatically converted into water-soluble monomers such as amino acids, glycerin, fatty acids, and monosaccharides and could thus be further degraded and transformed. In the second step, the products of hydrolysis are converted into acetic acid, hydrogen, carbon dioxide, organic acids, and alcohols. Some of these intermediate products can be used directly by methanogenic bacteria (Table 5.2). Only these products, as well as methanol, methylamine, and formate can be transformed by methanogenic bacteria into methane during the third step of methane formation.

The three stages of the anaerobic process of organic matter fermentation

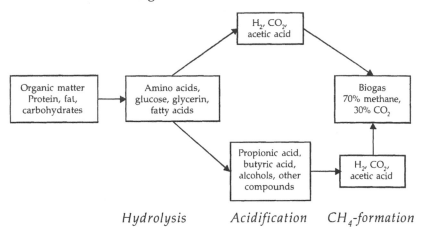

Hydrolysis   Acidification   CH$_4$-formation

Fig. 5.1: Three stages of the anaerobic process of organic matter fermentation (Sahm, 1981).

Table 5.2: Substrates for methanogenic organisms (Wolin and Miller, 1987; Knowles, 1993).

Substrates	Reaction	Products
$4H_2 + CO_2$	$2H_2O + CH_4$	$CH_4 + 2\ H_2O$
$4HCOO^- + 4\ H^-$	$4HCOOH$	$CH_4 + 3CO_2 + 2\ H_2O$
Methanol	$4CH_3OH$	$3\ CH_4 + CO_2 + 2\ H_2O$
Methanol + H_2	CH_3OH	$CH_4 + H_2O$
Aminomethane, other amines	$CH_3NH_2 + 2\ H_2O$	$3\ CH_4 + CO_2 + 4\ NH_4^-$
Acetate	$CH_3COO^- + H^-$	$CH_4 + CO_2$

About 70% of CH$_4$ emissions arise from anthropogenic sources and about 30% from natural sources (Table 5.3). Agriculture is considered to be responsible for 60 to 70% of the anthropogenic sources, with rice paddies and enteric fermentation and anaerobic waste processing being the major sources. The greatest sinks of CH$_4$ are reactions with OH° radicals in the troposphere (Crutzen and Andreae, 1991; Fung et al., 1991).

Table 5.3: Global estimated sources and sinks of methane (Watson et al., 1992).

Sources	$Tg\ CH_4\ yr^{-1}$
Natural	
Wetlands	100 to 200
Termites	10 to 50
Oceans	5 to 20
Fresh-water	1 to 25
CH_4 hydrate	0 to 5
Anthropogenic	
Coal mining, natural gas, and petrol industry	70 to 120
Rice paddies	20 to 150
Enteric fermentation	65 to 100
Animal wastes	10 to 30
Domestic sewage treatment	25
Land fills	20 to 70
Biomass burning	20 to 80
Sinks	
Atmopheric (tropospheric plus stratospheric removal)	420 to 520
Removal by soils	15 to 45
Atmospheric increase	28 to 37

Calculations (Mosier et al., 1998) place biological consumption of CH_4 as the largest global sink, exceeding even atmospheric oxidation. Microbial oxidation in soil and aquatic environments amounts to about 700 Tg yr^{-1}, which is 200 Tg yr^{-1} larger than about 500 Tg yr^{-1} emitted to the atmosphere (Table 5.3).

Knowles (1993) described CH_4 oxidation by methanotrophic organisms, which are obligate aerobes because the enzyme responsible for the initial step of CH_4 oxidation is a monooxygenase. This enzyme also has the ability to cooxidize ammonia, and apparently CH_4 and NH_3 are competitive substrates for both enzymes (Knowles, 1993).

Cellulose and lignin are structures of plants, and both are the most abundant biopolymers on earth and essential ingredients in city wastes (Table 5.1). Cellulose degradation occurs in anaerobic environments such as the rumen of ruminants, in swamps and in anaerobic digesters (Fig. 5.2). During hydrolysis the most water-insoluble biopolymers such as carbohydrates, proteins, and fats are decomposed by extracellular-enzymes to water-soluble monomers such as amino acids, fatty acids, and monosaccharides, and thus made accessible to further degradation. This step is inhibited by lignocelluloses, which are degraded only slowly or incompletely.

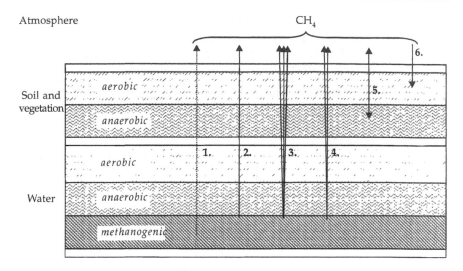

Atmosphere   CH$_4$

Soil and vegetation

aerobic

anaerobic

Water

aerobic   1.   2.   3.   4.

anaerobic

methanogenic

5.

6.

1. Diffusion; almost complete oxidation
2. Diffusion; oxidation already in O$_2$ containing sediment
3. Evolving gas bubbles that are not oxidized
4. Transport in plants; some oxidation in rhizosphere
5. Methane formation during flooding; some oxidation
6. Mainly oxidation

CH$_4$ Oxidation

Methylomonas, Methylococcus, Methylosinus

$$CH_4 \rightarrow CH_3OH \rightarrow$$
$$HCOOH \rightarrow CO_2$$

Fig. 5.2: Cellulose degradation in anaerobic environments (Haider, 1996).

For the anaerobic digestion of the organic fraction of municipal solid waste an average biogas yield of about 100 m^3 t^{-1} with a methane content of about 60% v/v can be assumed (Table 5.4).

Table 5.4: Gas yields from different municipal solid waste materials (Rilling et al., 1996).

Substance	Gas yield [m^3 kg^{-1} dry matter]	CH$_4$ content [vol.%]	CO$_2$ content [vol.%]
Carbohydrates	0.79	50	50
Fat	1.27	68	32
Protein	0.70	71	29
Municipal solid waste	0.1 to 0.2	55 to 65	35 to 45
Biowaste	0.2 to 0.3	55 to 65	35 to 45
Sewage sludge	0.2 to 0.4	60 to70	30 to 40
Manure	0.1 to 0.3	60 to 65	35 to 40

REFERENCES

Baffi C, Dell'Abate MT, Nassisi A, Silva S, Benedetti, A. Genevini,PL, Adani F, 2007: Determination of biological stability in composts. A comparison of methodologies. Soil Biol. Biochem. 39, 1284–1293.

Bollen GI, 1993: Factors involved in inactivation of plant pathogens during composting of crop residues. In: Science and Engineering of Composting, H.A.J. Hoiting, H.M. Keener (eds.), 301–318.

Bollen GI, Volkers D, 1996: Phytohygienic aspects of composting. In: The Science of Composting. de Bertoldi, M. Sequi, P. Lemmes, B. Papi (eds), Blackie Academic, London, pp. 233–246.

Crawford DI, 1988: Biodegradation of agricultural and urban waste. In: Actinomycetes in Biotechnology, M. Goodfellow and Smordarsky, eds. Acad. Press, San Diego, USA, pp. 433–499.

Crutzen PJ, Brühl C, 2001: Catalysis by NOx as the main cause of the spring to fall stratospheric ozone decline in the Northern Hemisphere. J. Phys. Chem. 105 A, 1589–1582.

Dick WA, Mc Coy EL, 1993: Enhancing soil fertility by addition of composts. In: H.A.J. Holtink and H.M. Keener (eds.): Science and Engineering of Composting, pp. 622–644.

Epstein E, Epstein JI, 1989: Public health issues and composting. Bio Cycle 30, 50–53.

Fernando T, Bumpus JA, Aust SI, 1990: Biodegradation of TNT (2,4.6 – trinitrotoluene) by *Phanerochaete chrysosporium*. Appl. Environm. Microbiol. 56, 1666–1671.

Fischbeck, G, Dennert, J, Müller, R, 1993: Investigations on supplemental N-fertilizer application for the optimum course of N-uptake into winter-wheat stands. *J. Agron. Crop Sci.* 171, 82–95.

Fung I, John J, LernerR J, Matthews E, Prather M, Steele LP, Fraser PJ, 1991: 3-Dimensional model synthesis of the global methane cycle. J. Geophys. Res. Atmosph. 96, 13033–13065.

Haider K, 1999: Microbe-Soil-Organic Contaminant Interactions.In: H.D. Skipper and R.F. Turco (eds.), Bioremediation of Contaminated Soils. Chapter 3, Agronomy Monographs 37. Amer. Soc. Agron. Madison WI, pp. 33–51.

Haider KM, Martin JP, 1988: Mineralization of C-14 labelled humic acid and of humic-acid bound C-14 xenobiotics by *Phanerochaete chrysosporium*. Soil Biol. Biochem. 20, 825–829.

Haug RT, 1993: The Practical Handbook of Compost Engineering, 717 pp., Fl Lewis Publ, CRC-Press, Boca Raton.

Kästner M, Mahro B, 1969: Microbial degradation of polycyclic aromatic hydrocarbons in soils affected by the organic matrix of compost. Appl. Microbiol. Biotechnol. 44, 668–675.

Kirk TK, Farell RL, 1987: Enzymatic combustion: The microbial degradation of lignin. Ann. Rev. Microbiol. 41, 465–505

Knowles R, 1993: Methane, Processes of production and consumption. In: L.A. Harper et al. (eds.), ASA Spec. Publ. No.55, pp145–156. Am. Soc Agron., Madison WI.

Knowles R, 1993: Methane. Process of production and consumption. In: Agricultural Ecosystem Effects on Trace Gas and Global Climate Change. L.A. Harper et al. eds. ASA Spec. Publ. 55, pp. 145–146.

Kutzner HJ, 2000: Microbiology of composting. In: Biotechnogy 11c, H.J. Rehm and G. Reed eds., pp. 36–100. Wiley-VCH, Weinheim.

Mathur SP, Owen G, Dinel H, Schnitzer M,1993: Determination of compost biomaturity. I Literature review. Biol. Agric. Hortic. 10, 65–85.

Millner PD, Olenchock SA, Epstein E, Rylander MD, Haines R, 1994: Bioaerosols associated with composting facilities. Compost Sci. Util. 2, 6–5.

Mosier AR, Duxbury JM, Freney JR, Heinemeyer O, Minami K, Johnson DE, 1998: Mitigating agricultural emission of methane. Climatic Change 40, 39–80.

Namkoong W, Hwang E, Park J, Choi J, 2002: Bioremediation of diesel-contaminated soil with composting. Environ. Poll. 119, 23–31.

Qicker P, Faulstich M, 2007: Technological aspects of energy from biomass. In: Round table discussion: Energy from biomass, Comm. Ecology, Bavarian Akad. Science, 33: 39–58, F Pfeil, Munich.

Rilling N, Arndt M, Stegmann R, 1996: Anaerobic fermentation of Biowaste at high total solid content—experiences with ATF-system. In: Management of urban biodegradable wastes. J.A. Hansen ed., Earthscan, pp. 172–180.

Rynk R, Richard TL, 2001: Commercial compost production systems. In: P.J. Stoffella and B.A. Kahn (eds). Compost utilization in horticultural cropping systems. pp. 51–93. Lewis Publ., Boca Raton FL.

Sahm H, 1981: Biologie der Methanbildung. Chem. Ing. Technol. 53, 854–863.

Scelza R, Rao MA, Gianfreda L, 2007: Effects of compost and of bacterial cells on the decontamination and the chemical and biological properties of an agricultural soil artificially contaminated with phenanthrene. Soil Biol. Biochem. 39, 1303–1317.

Scheibner K, Hofrichter M, Herre A, Michels J, Fritsche W, 1997: Screening for fungi intensively mineralizing 2,4,6-trinitrotoluene. Appl. Environ. Microbiol Biotechnol. 47, 452–457.

Steffen KT, Hattaka A, Hofrichter A, 2002: Degradation of humic acids by the litter decomposing basidiomycete Collybia dryophila. Appl. Environm.Microbiol 66, 3442–3448.

Waksman SA, 1926: The origin and nature of the soil organic matter or soil "humus". Soil Sci. 22: 123–162; 232–333; 395–406; 421–436.

Waksman SA, Umbreit WW, Gordon TC, 1931: Thermophilic actinomycetes and fungi in scils and in composts. Soil Sci. 47: 37–61.

Watson RT, Meira LG, Sanhuueza E, Jaanetos T, 1992: Greenhouse gases: Sources and Sinks. In: Houghton et al. Eds. The Suppl. Rep. To IPCC. Scientific Assessment. Cambridge Univ. Press pp. 25–46.

Widmann B, 2007: Biomass for production of heat, fuel and electricity. Round table discussion: Energy from biomass, Comm. Ecology, Bavarian Akad. Science, 33: 27–38, F Pfeil, Munich.

Wolin MJ, Miller TL, 1987: Bioconversion of organic carbon to CH_4 and CO_2. Geomicrobiol J. 5: 239–259.

Wyk van JPH, 2001: Biotechnogy and the utilization of biowaste as a resource for bioproduct development, Review. Trends Biotechnol. 19, 172–177.

Trace Gases in Soil

Composition of the atmosphere

The todays composition of the atmosphere results preponderantly from biological activity, e.g., N_2 by the denitrification of nitrate. Without continuous regeneration of O_2 and N_2, CO_2 should continuously increase.

Beside the main components O_2 (20.9%) and N_2 (78.1%) the atmosphere contains noble gases (<1%) and also small amounts of other gases such as CO_2, CH_4, NH_3, N_2, CO, NO_x and others in the range of ppmv (1 μL L^{-1}) and ppbv (1 nL L^{-1}). The important function of some of these trace gases in regulating climate turns them nowadays more and more into the center of interest.

For their formation and degradation soils and their organisms as well as plants have a decisive role. Their additional release by agricultural and industrial activities disturbs increasingly the natural equilibrium between production and degradation. This leads to a more or less drastic increase of their atmospheric concentration of 0.3% to more than 1% per year (Fig 6.1).

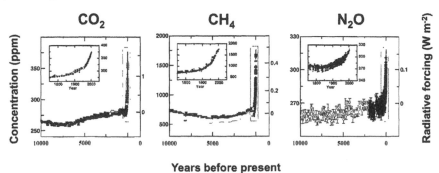

Years before present

Fig. 6.1: Increase of atmospheric concentration of the trace gases CO_2, CH_4 and N_2O and radiative forcing (IPCC, 2007).

Production and degradation of atmospheric trace gases

In many cases the flow rates of trace gases from distinct sections are not fully known and therefore ranges given by different authors are quite variable (Table 6.1; see also Table 4.13).

Table 6.1: Estimations of annual production and degradation of the atmospheric trace gases CO_2, CH_4, N_2O. (Houghton 1992; Duxberry et al. 1993; according to Haider 1996).

CO_2	CH_4	N_2O
736 * 10^{15} g C	3.6 to 3.8 * 10^{15} g C	1.5 * 10 ^{15} g N

Annual sources or sinks of trace gases by different processes are given in Table 6.2 (Houghton 1992; Duxberry et al., 1993; according to Haider, 1996).

Table 6.2: Sources and sinks of the atmospheric trace gases carbon dioxide, methane and nitrous oxide.

		$CO_2[10^{15}$ g C a$^{-1}]$		$CH_4[10^{12}$ g C a$^{-1}]$		$N_2O[10^{12}$ g N a$^{-1}]$
Sources						
natural	Plant respiration	60	Wetlands, swamps	85	Oceans	1.4–2.6
			Oceans, lakes	10		
	Soil degradation	60	Game, termites	30	Tropical forests	2.2–3.7
	Release from oceans	105	Forest and prairie fires	20	Savannah	0.5–2.0
					Forests	0.1–2.0
Subsum		225		145		4.2–10.3
Anthropogenic	Fuel combustion	5.9	Rice culture	85	Cultivated land	0.03–2.0
	Soil degradation by agriculture	1.8	Domestic animals	60	Combustion of biomass	0.3–1.3
			Landfills	30	Production of nylon and HNO$_3$	0.5–0.9
Subsum		7.7		255		0.83–4.2
Sum sources		232.7		400		5–14.5
Anthropogenic contribution (%)		3.3		64		17–29
Sinks						
	Uptake in oceans	106.7	Reaction with OH radicals	375	Photolysis in stratosphere	7–13
	Photosynthesis	120	Oxidation in soils	26		
Sum sinks		226.7		401		7–13

The difference between sources and sinks is insecure; atmospheric measurements showed that after the industrial revolution there was an increase in global atmospheric methane concentrations. In the past 200 years methane concentrations have increased from approximately 620 to approximately 1700 ppb.

Greenhouse effect and soil environment

Dickinson and Cicerone (1986) called the influence of trace gases such as methane and others on radiation (Fig. 6.2) as thermal trapping, a process leading to an increased green house effect with increasing trace gas concentrations (Solomon et al., 2007).

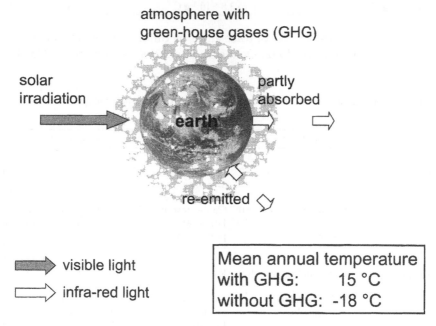

Fig. 6.2: Principle of trace gases as "thermal trapping" of heat radiation from earths surface (green house effect).

Attention of gases other than CO_2 continued to grow, and in this respect methane offers some peculiar possibilities by its increased emissions from wetlands and especially by its huge reservoirs in the permanently frozen peat often many meters deep. Layers underlying arctic tundra emit even more methane and equal the present amounts of carbon dioxide presently in the atmosphere. Each molecule of methane has a green house effect of more than 20 times of that of a molecule of carbon dioxide.

The current greenhouse effect of N_2O is not very large (5%) of the present trace gas warming potential. However, since this gas remains in the atmosphere for a century or more and due to the emissions from fertilizer and manures N_2O will become nearly as important as CO_2 or methane in the future (see Table 6.3).

Meanwhile, the actual rate of increase of methane is 1% a year bringing a shocking 11% increase of methane in the past decade. Alarming are the enormous quantities of methane locked in "clathrates", i.e. methane hydrates, found in the muck of sea beds around the world. Clathrates are ice like substances with methane imprisoned within their structure. It was pointed out that if a slight warming penetrates the sediments, the clathrates might melt and release colossal bursts of methane and CO_2 into the atmosphere.

Table 6.3: Present atmospheric concentrations and trace gases: (residence times and procentual contributions to the green house effects).

	Concentration in ppmv	Residence time in years[1]	Annual increase % per year	Relative greenhouse potential[2]	% contribution to the present greenhouse effects[3]
CO_2	354	120	0.5	1	ca. 50
CH_4	1.72	10	1.1	21	ca. 19
N_2O	0.32	150	0.25	206	ca. 5

1. Mean residence time of a gas in the atmosphere.
2. Percentage contribution of a gas to the respective increase of the green house effect.
3. The difference to 100% is given by participation of ozone, fluorochlorohydrocarbons and water vapour.

Quantifying of trace gases

The two non-CO_2 greenhouse gases nitrous oxide and methane are mainly generated from mineral N and by farm animals. A number of methods is available to quantify their emissions and their possible mitigation. They are used to measure rates of exchange of CH_4 and N_2O between the soil surface and the atmosphere. They include simple and widely used enclosure methods (chambers) and micro-meteorological methods with various degree of complexity (Eddy covariance, flux gradients) (Denmead, 2007). Measurements for their quantification are made at regular intervals after temporarily sealing the chambers with an air tight seal.

Chambers are cylinders or boxes randomly inserted into the soil to form an air-tight soil enclosure. Alternatively intact soil cores can be collected in metal cylinders and gas fluxes are monitored in a gas chromatograph (Hedley, 2002) (Fig. 6.3). The capacity of soils to oxidize CH_4 appears possible for measurements that can be made to study e.g. the influence of grazing animals and the influence of climatic variables regulating emissions (Ojima et al.,1992).

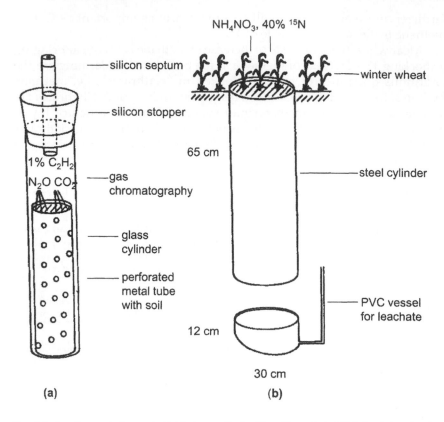

Fig. 6.3: A soil sample in an air tight glass cylinder (Benckiser et al., 1986) for determination of the N_2O release. (a) Soil cylinders with lysimeter bottles for determination of gaseous N– and leaching-losses. (b) Soil cores in glass cylinders for determination of the N_2O release.

Some studies show that ammonium N (NH_4^+–N) can reduce the capacity to oxidize methane, since both reactions are catalyzed by monooxygenases.

Trace gas transport and emissions within soils

The pathways for transport and exchange in the interior of soil and the atmosphere are regulated by the following parameters.

1. diffusion/advection/convection through the air-filled pores.
2. diffusion through liquid water layers.
3. bubble transport through liquid water layers.
4. vascular transport within plants.
5. pumping by changes in atmospheric pressure.

Models have been developed for diffusion fluxes including microbial production and consumption of trace gases by Galbally and Johansson (1989) of NO and other gases. Applying such model the residence time of biogenic trace gases in soils can be estimated.

Soils as sources and sinks of methane

Already in soil during methane production and its diffusion through aerated soil layers a significant biological oxidation of methane takes place by methane oxidizing bacteria which need O_2 (Conrad, 1989). Several species of *Methylomonas*, *Methylococcus*, *Methylosinus* and others use methane by oxidation by a Cu-dependent cytochrome P450 monooxygenase to methanol and finally CO_2.

$$CH_4 \rightarrow CH_3OH \rightarrow HCHO \rightarrow HCOOH \rightarrow CO_2$$

The cells obtain most of their energy and building blocks from these intermediary C_1 compounds (Schlegel, 1993). Most methanotrophic organisms are mesophilic, but several are also thermophilic. But even in soils of the northern area under permafrost conditions, where large amounts of methane are produced, no psychrophilic species adopted to colder climate with growth optima below 20°C could be isolated.

The highly diluted atmospheric methane concentrations cannot be oxidized by pure cultures of methanotrophic organisms, but in soils there may be an enrichment of methane by sorptive processes. Substrates of methanogenic organisms are summarized in Table 6.4. Any methane oxidizing organisms need O_2 and cannot use any other electron acceptors such as nitrate.

Table 6.4: Substrates for methanogenic organisms
(Wolin and Miller, 1987; Knowles, 1993).

Substrate	Reaction		Product
H_2 and CO_2	$4 H_2 + CO_2$	$\rightarrow$	$CH_4 + 2 H_2O$
Formiate	$4 HCOOH$	$\rightarrow$	$CH_4 + 3 CO_2 + 2 H_2O$
Methanol	$4 CH_3OH$	$\rightarrow$	$3 CH_4 + CO_2 + 2 H_2O$
Methanol + H_2	$CH_3OH + H_2$	$\rightarrow$	$CH_4 + H_2O$
Methylamine and other amines	$4 CH_3NH_3^- + 2H_2O$	$\rightarrow$	$CH_4 + CO_2$
Acetate	CH_3COOH	$\rightarrow$	$CH_4 + CO_2$

In arctic soils vascular bundles similar as in rice or in paddy soils are adapted to the transport of methane. Here Carex or Eriophorum vegetation can transport 50 to 100% of the methane to the soil surface (Reeburgh and Whalen, 1992; see Fig. 5.2).

Also green plants are a source of methane production (Fig. 5.2). The amount of methane produced by plants has not been quantified yet but first estimations point to a significant contribution of terrestric plants to the global methane production using pectine and other unknown plant constituents as a source under aerobic conditions and UV light exposure (Keppler et al., 2008).

NH_3 probably inhibits competitively the oxidation of methane in soils since both substrates are oxidized by monooxygenases (Hütsch and Webster, 1993; Mosier et al., 1991). Products formed from ammonia are NH_2OH and nitrite. Since nitrifying and methane oxidizing organisms occur in a similar environment they both compete for O_2, CH_4 and NH_3 (Bédard and Knowles, 1989).

Tate (2001) found that some forest soils contain active methanotrophs that consume CH_4 rapidly, but yet unknown are the opportunities for enhancing oxidation rates.

Strategies for reducing N_2O emission

Options have been suggested by Bolan et al. (2004) and Granli and Böckman (1994). They include improvement of the overall N-use efficiency, lower N-content of pastures, supplementary feed with lower N contents in order to reduce N excretion, the use of controlled low nitrogen doses as fertilizer, and the reduction the livestock numbers per area.

Microbial processes as sources of N_2O in natural systems

A further trace gas preponderantly originating from biological sources is nitrous oxide. Its concentration increased during the last 100 years from 280 to 310 ppb whith a further annual increase by 0.25%. The atmosphere contains nowadays about $1.5 \cdot 10^9$ tons of N in form of N_2O. This is equivalent to an increase of $3–4\ 10^6$ tons N_2O nitrogen per year. It can be suspected that 80 to 90% of the emission originates from soils (Bouwman, 1990) where it is formed by nitrification and denitrification (see chapter 4).

Several biological activities lead to the formation of N_2O:

(a) chemolithotrophic (autotrophic) nitrification and
(b) denitrification,
(c) heterotrophic nitrification by fungi and bacteria.

Agricultural management and release of N_2O

The N_2O release from soils depends significantly on the soil water content and the redox potential in the soil environment. At low water content there is a low contribution of nitrification increasing at higher water contents (see Fig. 4.11). At high water contents the N_2O release is more and more coined by denitrification. Nitrous oxide is a very long living nitrogen gas and increases due to intense worldwide application of

N-fertilizer. It is estimated that 70–80% of the N_2O release is caused by anthropogenic activities. Nitrogen compounds do not accumulate permanently in soils, sea or sediments, since any applied N-compounds become ultimately nitrified and denitrified, careless whether it is applied as organic or inorganic fertilizers. Estimated 2–3% of these nitrified or denitrified fertilizers are released as N_2O which explains the annual increase of 2.4–3.7 Mt N a^{-1}.

Nitrous oxide can be consumed by soils especially in highly anaerobic wetlands where N_2O is readily consumed as an electron acceptor by denitrifiers.

It is clear that under urea- or ammonium-fertilized conditions nitrification is the major source of N_2O (Bremner and Blackmer, 1981), but there is general agreement that denitrification is the major source of N_2O and possibly of NO. Other potential sources for N_2O are heterotrophic nitrification performed by a broad spectrum of bacteria and fungi.

Changes in soil conditions such as by freeze/thaw events, wet/dry cycles and seasons alter N_2O production and release from soil to the atmosphere (Goldberg et al., 2008).

Emission of NH_3

Agricultural sources contribute annually about 40 Mt of N as NH_3 and NOx which equals 20–25% of the mineral fertilizer application (see Fig. 4.6). According to Isermann (1992) the annual entry of NH_3 and NO in European areas amounts to 10–80 kg N ha^{-1} or more. NH_3 and NO do not have any influence as trace gases on the climate but have a strong impact on forests by overfertilization and acidification in other N-limited area (Table 4.5). Measurements in a woodland area of the Netherlands resulted in depositions of 17 kg N $ha^{-1} a^{-1}$ with the rain and of 42 kg N $ha^{-1} a^{-1}$ by runoff from leaves (van Breemen et al., 1987).

Further informations of global sources and sinks of ammonia is given in Table 4.5 and Fig. 6.4 (Schlesinger and Hartley, 1992).

Soils and turnover of organic carbon

Soil organic matter globally contains about $1,500 \times 10^9$ tons carbon and thus more than the carbon in the atmosphere (ca. 750×10^9 tons carbon) (Fig. 6.5). Temperature increase can accelerate humus degradation to CO_2 which would lead to a tremendous increase in atmospheric CO_2 content, one of the main climate relevant trace gases.

The atmospheric increase of CO_2 between 1800–1950 resulted mainly from the conversion of forest- and prairie soils into agricultural land. The later on observed increase to now 357 ppmv is mainly caused by combustion of fossil fuels and by increased degradation of SOM at

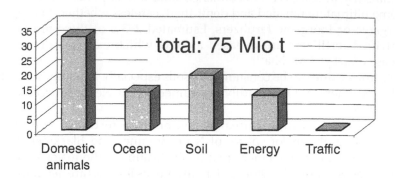

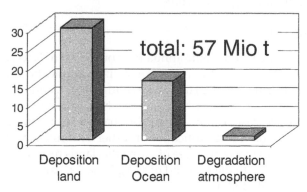

Fig. 6.4: Global estimates of NH₃ emission and deposition
(Schlesinger and Hartley, 1992).

temperature increase, and by fuel clearing of natural forests (30%). Predictions of a future CO_2 increase prognose a further significant increase for the next hundred years. Such an increase causes simultaneously an increase in temperature and degradation of the humus reserves presently occurring in the worldwide soils (Fig. 3.3).

In case that the mean annual temperature (MAT) will increase in the next 50 years by about 3° C the additional CO_2 release from SOM will amount to 100×10^{15} g C (Table 6.5).

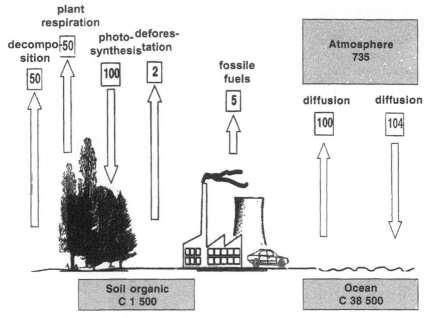

Fig. 6.5: Global cycle of carbon in 10^9 t C per year.

Table 6.5: Estimation of CO_2-C-release from worldwide humus reserves by temperature increase during the next 60 years of 1.2; 1.8 and 3° C (according to Jenkinson et al., 1991).

Temperature increase in the next 60 years (°C) [a]	Additional CO_2-release from SOM	C in SOM	Simultaneous C release by combustion of fossil fuel	Present atmospheric CO_2 content
	10^{15} g	%	%	%
1.2	41	2.7	12.7	5.3
1.8	61	4.1	18.8	7.8
3.0	100	6.7	30.9	12.8

a. Increase of the mean annual temperature at soil surface.

A global temperature increase certainly should speed dramatically the degradation, mainly that of plant residues and that of more easily humus components (Table 3.5). This increase however should not lead to a drastic degradation of the great humus stocks. The degradation should foremost diminish the less humified compounds which are responsible for the nutrient- and C-flux in soil and forest soil structure. It also should lead to an increase in soil fertility by better plant growth (Cast Report, 1992, Wittwer, 1988).

An increase in atmospheric temperature together with a lowering of the water table will increase the rate of carbon dioxide efflux from the tundra (Billings et al., 1983).

Ultimately, the net effect of climate change on ecosystems carbon budget depends on the balance between photosynthesis and respiration (autotrophic root respiration and heterotrophic microbial respiration). There is however a considerable lack in our understanding of the response of soil respiration (Trumbore, 2006).

In fact, climate change can have both direct and indirect effects on the activities of soil microbes, but there is a prime uncertainty relating the relationship between the temperature sensitivity of more recalcitrant substrates and that of more labile substrates.

It is well established that elevated CO_2 increases plant photosynthesis and growth and this in turn leads to an enhanced flux of carbon to roots. This may lead to an increased formation of recent organic carbon but also to enhance degradation of old soil organic carbon, a phenomenon known as priming effect (Kuzyakov, 2006). Our understanding of the involved complicated feedback mechanisms is rather poor (Wardle et al., 2004; Bardgett et al., 2008).

An estimate in the global emission of green house gases and the contribution of different sources is shown in Fig. 6.6.

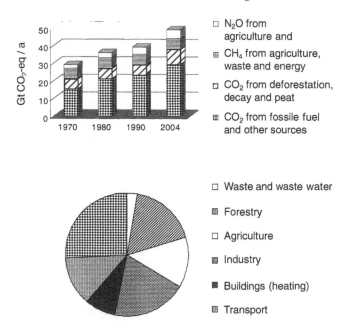

Fig. 6.6: (Top) Global anthropogenic emissions of green house gases. (Bottom) Share of different sectors in green house gas emission. IPCC, 2007.

REFERENCES

Bédard C, Knowles R, 1989: Physiology, biochemistry, and specific inhibitors of CH_4, NH_4^+, and CO oxidation by methanotrophs and nitrifiers. Microbiol. Rev. 53, 68–84.

Bardgett RD, Freeman C, Ostle NJ, 2008: Microbial contributions to climate change through carbon cycle feedbacks. ISME J. 2, 805–814

Benckiser G, Haider K, Sauerbeck D, 1986: Field measurements of gaseous nitrogen losses from an Alfisol planted with sugra-beets. Z. Pflanzenernähr. Bodenk. 149, 249–261.

Billings WD, Luken JO, Mortensen DA, Peterson KM, 1983: Increasing atmospheric carbon dioxide: possible effects on arctic tundra. Oecologia 58:286–269.

Bolan NS, Saggar S, Luo JF, Bhandral R, Singh J, 2004: Gaseous emissions of nitrogen from grazed pastures: Processes, measurements and modelling, environmental implications, and mitigation. Adv. Agron. 84, 37–120.

Bouwman AF, 1990: Exchange of green house gasses between terrestrial ecosystems and the atmosphere. In Soils and the Greenhouse Effect, Bowman AF ed., Wiley and Sons, New York, pp 61–127.

Bremner JM, Blackmer AM, 1981: Terrestrial nitrification as a source of atmospheric nitrous oxide. In: Denitrification, nitrification and atmospheric nitrous oxide. Delwiche CC ed., Wiley & Sons, New York, 151–170.

CAST Report 119, June 1992: Preparing U.S. Agriculture for Clmate Change. Ames, Iowa, 96 pp.

Conrad R, 1989: Control of methane production in terrestrial eco-systems and the atmosphere. In: Exchange of Trace Gases between terrestrial Ecosystems and the atmosphere. Andreae MO . Schimel DS eds. Wiley & Sons, Chichester. p 39–58.

Denmead OT, 2007: Approaches to measure fluxes of trace gases between landscapes and the atmosphere. Plant Soil-Special Issue 26: on CO3 flux research 25, 42.

Dickinson RE, Cicerone RJ, 1986: Trace gases update. Environment 28, 2–3.

Duxberry JM, Harper JA, Mosier AR, 1993: Contributions of ecosystems to global climate change. In: Agricultural eco-systems effects on trace gases and global climate change. Harper LA et al. eds. ASA Spec. Publ. No 55, 1–18

Galbally IE, Johansson C, 1989: A model relating laboratory measurements of rates of nitric-oxide production and field-measurements of nitric oxide emissions from soils. J. Geophys. Res.—Atmospheres 94, 6473–6480.

Goldberg SD, Muhr J, Borken W, Gebauer G, 2008: Fluxes of climate-relevant trace gases between a Norway spruce forest soil and atmosphere during repeated freeze-thaw cycles in mesocosms. J. Plant Nutr. Soil Sci. 171, 729–739.

Granli T, Böckman OC, 1994: Nitrous oxide from adriculture. Norweg. J. Agric. Sci. Suppl. 12, 1–27.

Haider K, 1996: Biochemie des Bodens, Enke Stuttgart. 174 pp.

Hedley CB, Saggar S, Tate KR, 2002: Procedure for fast simultaneous analysis of the greenhouse gases: methane, carbon dioxide, and nitrous oxide in air samples. Comm. Soil Sci. Plant Anal. 37, 1501–1510.

Houghton IT, Callander TA, Varney 1992 SK, eds. Climate change 1992. The supplementary report to the IPCC scientific assessment. Cambridge Univ. Press 200 pp.

Hütsch B, Webster CP, 1993: Effect of nitrogen fertilization on methane oxidation in the Broadbalk wheat experiment. Mitt. Bodenkundl. Ges. 69, 227–230.

IPCC, 2007: Summary for Policymakers. In: Climate Change 2007: The Physical Science Basis. Contribution of Working Group 1 to the Fourth Assessment Report of the Intergovernmental Panel on Climate Change [Solomon S, Qin D, Manning M, Chen Z, Marquis M, Averyt KB, Tignor M, Miller HL (eds.)] Cambridge University Press, Cambridge, United Kingdom and New York, NY, USA.

Isermann K, 1992: Territorial, continental and global aspects of C,N, P and S emissions from agicultural ecosystems. In: NATO Advanced Research Workshop on Interactions of C, N, P and S Biochemical Cycles. Springer, Heidelberg.

Jenkinson DS, Adams DE, Wild A, 1991: Model estimates of CO2 emissions from soil in response to global warming. Nature 351, 304–306.

Keppler F, Hamilton JTG, McRoberts WC, Vigano I, Braß M, Röckmann T, 2008: Methoxyl groups of plant pectin as a precursor of atmospheric methane: evidence from deuterium labelling studies. New Phytologist 178, 808–814.

Knowles R, 1993: Methane: Process of production and consumption. In: Agricultural Ecosystem Effects on Trace gas and global climate change. ASA Spec. Publ. 55, 145–156.

Kuzyakov Y, 2006: Sources of CO_2 efflux from soil and review of partitionig methods. Soil Biol. Biochem: 38: 425– 448.

Mosier AR, Schimel D, Valentine D, Bronson K, Parton W, 1991: Methane and nitrous oxide flux in native, fertilized and cultivated grasslands. Nature 350, 330–332.

Ojima DS, Svensson BH, eds., 1992: Trace gas exchange in a global perspective. Ecol. Bull. (Copenhagen) 42, 1–206.

Reeburgh WS, Whalen SC, 1992: High latitude ecosystems as CH_4 sources. Ecol. Bull. 42, 62–70.

Schlegel HG, 1993. Allgemeine Mikrobiologie , 7. edit. Thieme, Stuttgart, 634 pp.

Schlesinger WH, Hartley AE, 1992: A global budget for atmospheric ammonia. Biogeochem. 15, 191–211.

Solomon S, Qin D, Manning M, Chen Z, Marquis M, Averyt KB, Tignor M, Miller HL (eds.), Climate change 2007: The Physical Science Basis. Contribution of Working Group I to the Fourth Assessment Report of the Intergovernmental Panel on Climate Change (IPCC). Cambridge University Press, Cambridge, United Kingdom and New York, NY, USA, pp. 499–587.

Tate RL, 2001: Soil organic matter: Evolving concepts. Soil Sci. 166, 721–722.

Trumbore S, 2006: Carbon respired by terrestrial ecosystems—recent progress and challenges. Glob. Change Biol. 12, 141–153.

van Breemen N, Mulder J van Grinsven :, 1987: Impacts of acid atmospheric deposition on woodland soils in the Netherlands. II. Nitrogen transformation. Soil Sci. Soc. Am. J. 51 1634– 640.

Wardle, 2004: Vulnerability to global change of ecosystem goods and services driven by soil biota. Sust. Biodiv. Ecosyst. Serv. Soils Sedim. 64, 101–135.

Wittwer SH, 1988: The greenhouse effect. Carolina Biol. Readers 163: 1–15

Wolin MJ, Miller TL, 1987: Bioconversion of organic carbon to CH_4 and CO_2. Geomicrobiol. J. 5, 239–259.

7

Heavy Metals as Pollutants: Toxicity, Environmental Aspects, Resistance

Human activities have contaminated large areas in both developed and developing countries. The European Environment Agency has estimated the total costs for the clean-up of contaminated sites in Europe to be between EUR 59 and 109 billion. Heavy metals are a major factor of this pollution due to their atmospheric deposition, their leaching tendency, and the fact that they are undegradable.

Soils have a natural content of heavy metals in sometimes big amounts in the form of minerals and or oxides not easily available to microorganisms, plants or animals. Many of these minerals are essential trace elements for organisms and necessary for the function of their enzymes (Cu, Zn, Fe, Mn, Mo, Se, V, Co) or as growth factors (Zn, Cu and others). Other elements (Pb, Cd or Hg) have no function as trace elements but can be tolerated at low concentrations, but are toxic at higher concentrations. They are discharged from industry or domestic waste in the environment and in soils.

Heavy metals comprise about 65 metallic elements with densities greater than 5g cm^{-3}. They may have considerable impacts on microbial populations and their activities in soil. A reduction in biomass as well as microbial activities has been reported to be potentially affected by toxic metal exposure (Kuperman and Carreiro, 1997). In addition changes in microbial community structure in response to metal contamination may result in reduced abilities to degrade aromatic and phenolic compounds (Doelman et al., 1994; Reber, 1992). In general numbers of resistant organisms growing on substrates containing aromatic structures are lower than that of sensitive organisms. This indicates that organisms with resistance against zinc or cadmium limit their capacity for degradation of distinct substrates.

The global industrial age production of Cd, Cr, Cu, Hg, Ni, Pb, and Zn, and their potential accumulation and environmental effects caused that world soils have been seriously polluted by Pb and Cd and slightly by Zn. The potential anthropogenic heavy metal input into the pedosphere increased tremendously after 1950. The improvement of industrial processing technologies and the recycling of metal containing products and the development of new substitute materials are possible strategies to minimize the effect of heavy metals on our environment.

Essential and toxic elements

Some elements have been found to be essential for life, however ineffective at too low and toxic at excessive concentrations. As an example, copper is essential for higher plants, and algae and serves as primary electron donor in photosystems, and as cofactor in oxidases, mono-and di-oxygenases.

The form (speciation) of an element modifies its availability and the possibility to interact with an organism, i.e., as free cation (often the most toxic form), as complex with negative counterions (e.g. $Al(OH_4)^-$ or $CdCl_3^-$), or as complex with and without charge bound to organic or inorganic ligands.

Concern about the input of metals to terrestrial ecosystens is related to the:

(1) impact on soil organisms and plants,
(2) effects on soil organisms including microorganisms/microfungi and soil fauna such as nematodes and earthworms including biomass changes,
(3) impact on aquatic organisms due to runoff by surface and ground waters,
(4) uptake via food chain potentially affecting humans by reducing food quality of crops and animal products,
(5) impact on animal health through accumulation in organs of cattle, birds, mammals. This is specifically considered important with respect to cadmium and mercury and to a lesser extent to lead.

A risk assessment approach is to determine the maximum load that causes no or tolerable damage (long term acceptable load or critical load). Approaches to calculate critical loads is based on the balance of all relevant loads of heavy metals in a considered ecosystem (De Vries and Bakker (1998).

Long term effects on organisms

Soils containing toxic metals, e. g., Cu, Ni, Zn, Cd, from long term input of contaminated sewage sludge then may contain less biomass with altered microbial functionality. The general order of inhibition was found to be Zn < Cu < Cd for single metals (Giller, 1998). Experiments by metal addition to soils in the laboratory might have little relation to results obtained in field experiments (Giller et al., 1998).

Population shifts by metal addition from bacteria, including streptomyces, towards fungi were reported. This may last only a few days after metal exposure, but other studies demonstrated several years lasting effects (Doelman and Hanstra, 1979; Chander et al., 2001).

Microorganisms have developed mechanisms to cope a variety of toxic metals. Bacterial plasmide resistance genes to many toxic metals and metalloids (e.g. Ag^+, As^{3+}, Cd^{2+}, CrO_4^{2+}, Cu^{2+}, Zn^{2+}) are known. Multiple metal resistant transferable phenotypes in bacteria may thus serve as indicators of soil contamination with heavy metals (Ryan, 2005). Relatively little is known about the resistance against lead (Roane, 1999); see Fig. 7.1.

Resistance genes are frequently located on bacterial chromosomes. Plasmid metal resistance is highly specific. However, also other less specific interactions, e.g. sorption of heavy metals to clay minerals or humic substances, contribute to the overall response against heavy metals.

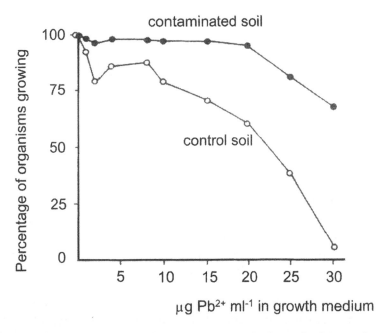

Fig. 7.1: Development of microbial resistance against lead on a lead contaminated soil after 2 years (Starkey and Waksman, 1943).

As with bacteria, in fungi the vacuole has an important role in regulating the cytosolic metal ion concentration. Metals are preferentially sequestered by the vacuole, e.g. Mn^{2+} (Gadd and Lawrence, 1996), Fe^{2+} (Bode et al., 1995) and other metal ions.

Cadmium (Cd^{2+}) is transported into the cells by a chromosomally encoded transport system. The resistant cells rapidly pump Cd^{2+} out and ATP is involved in this pumping mechanism.

As shown in Table 7.1 sewage sludge usually contained and contains significantly higher concentrations of heavy metals than uncontaminated soils. Toxic metals may be resorbed by plants from the soils depending on the respective soil plant transfer factors (Uchida et al., 2007). Uptake by the plants depends on the availability and mobility of the metals. Soils with a high pH and a high cation exchange capacity will immobilize most added metals. A drop in pH and could cause a release of the metals. Based on the results of Alloway (1968) shown in Table 7.2 limit values for metal concentrations in soils and sewage sludge were deducted. Such trigger values are regularly updated in different countries.

Table 7.1: Heavy metal composition of typical uncontaminated soils and in sewage sludge (Alloway, 1968; O'Neill, 1994; Alloway, 1999; Uchida et al., 2007).

Element	Average soil content, mg/kg	Average sewage sludge content, mg/kg	Mobility of ions	Transfer factors soil/plants*
Fe	30,000	16,000	Slightly mobile	$10^{-4} - 10^{-3}$
Cu	50	250	Slightly mobile	$10^{-3} - 10^{-1}$
Cr	50	500	Immobile	$10^{-4} - 10^{-2}$
Zn	100	3,000	Well mobile	$10^{-2} - 10^{0}$
Pb	25	700	Highly mobile	$10^{-4} - 10^{-2}$
Cd	0, 4	20	Well mobile	$10^{-2} - 10^{0}$
Hg	0, 25	2	Immobile	not determined

* uptake values from 68 different plants.

Metallothionein

Metallothioneins are cysteine-rich, low molecular weight proteins (3.5–14 kDa). They have the capacity to bind both essential (Zn, Cu, Se,...) and toxic (Cd, Hg, Ag,...) metals through the thiol group of its cysteine residues which represent about 30% of its amino acids (Fig. 7.2) (Kägi and Schäffer, 1988).

Metallothionein function is not clear, but they may provide protection against metal toxicity, be involved in regulation of physiological metals (Zn and Cu) and provide protection against oxidative stress. Human MT binds three Zn^{2+} in the beta domain, Zn_3Cys_9, and four Zn^{2+} in the alpha domain, Zn_4Cys_{11} (Messerle et al., 1992). By binding and releasing zinc, metallothioneins may regulate zinc level within the body.

Table 7.2: Importance of heavy metals for organisms (e essential, t toxic) and limit values in soil and sewage sludge (Alloway, 1968).

Element	Micro-organisms, plants	Animals, men	Limit value soil, mg/kg	Limit value sewage sludge, mg/kg
Fe	e	e	None	None
Cu	e, t	e, t	60	800
Cr	e, t	e, t	100	100
Zn	e, t	e	200	2500
Pb	t	t	100	900
Cd	t	t	10	1,5
Ni	t	e	50	200
Hg	t	t	1	8

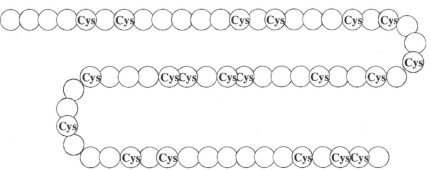

Fig. 7.2: Cystein distribution in the amino acid chain of metallothionein.

In mammals, large quantities are synthesised in the liver and kidneys. Their production is dependent on the availability of the dietary minerals (Zn, Cu, Se) and the amino acids histidine and cysteine.

Metallothioneins comprise three classes: Class I with homology with horse MT, Class II, with no homology with horse MT, and Class III, which includes phytochelatins, Cys-rich enzymatically synthesised glutathione-oligomers in plants which are no longer considered MTs.

Transfer functions

They describe the partitioning between metals in soil and soil solution and are used to calculate critical metal contents for different soil conditions. As an example, plants accumulate cadmium but there is only a loose connection with Cd-content in soil (Knoche et al., 1999). The ions can only be transferred if being solubilised (Table 7.1).

Mobilization

Microorganisms can mobilize metals through autotrophic and heterotrophic leaching, chelation by microbial metabolites and siderophores, and methylation which may result in volatilization (e.g., CH_3HgCH_3). All these processes can lead to dissolution of insoluble metal compounds and minerals including oxides, phosphates, sulphates and more complex mineral ores. Organic acids can supply both protons and complexing anions (Burgstaller and Schinner, 1993). For example citrate and oxalate can form stable complexes with many metals (Klees, 1993).

Most chemolithotrophic leaching is carried out by acidophilic bacteria which can fix CO_2 and obtain energy from the oxidation of Fe(II) or reduced sulphur compounds (Schippers and Sand, 1995). The microorganisms involved include sulphur oxiding bacteria e.g. *Thiobacillus thiooxidans* or *T. ferrooxidans* and others (Rawlings, 1997).

Phytoremediation

Phytoremediation is defined as the use of green plants to remove pollutants from the environment or to render them harmless (Raskin et al., 1997). It has long ago been noticed that some plants are able to accumulate high concentrations of metals. Such hyperaccumulators, by definition, accumulate at least 100 mg kg^{-1} (0.01% dry weight) Cd, As and some other trace metals, 1000 mg kg^{-1} Co, Cu, Cr, Ni and Pb and 10000 mg kg^{-1} Mn and Ni, respectively (Evangelou et al., 2007).

Phytoremediation studies have continued to focus on hyperaccumulating species such as *Thlaspi caerulescens*, *T. rotondifolium* and *Alyssum lesbiacum*. Until now over 400 plant species have been identified as natural metal hyperaccumulators, representing <0.2% of all angiosperms. A characteristic of these species is their slow growth and limited biomass production. As total metal extraction is the product of biomass and tissue concentration, the speed of metal removal is accordingly limited. Calculations showed that for instance, in the case of Pb, this technology can only be feasible if systems can be developed to employ high biomass plants, which are capable of accumulating more than 1% Pb in shoots and produce more than 20 t of biomass ha^{-1} yr^{-1}. Hyperaccumulating species cannot meet such requirements.

To increase the effectiveness of metal uptake, several approaches are possible: the addition of chelating agents will increase the bioavailability of metals in soil, thus enhancing plant uptake but also the leaching potential. Genetic engineering of plants by inserting genes coding for metal binding proteins, e.g. metallothionein, or peptides, e.g. phytochelatines, may help to increase the tolerance and uptake efficiency of plants.

The main advantage of phytoremediation is its low cost compared with conventional remediation methods such as excavation and reburial, even though taking into account the long-term strategy that must be applied to clean up soil effectively. It offers permanent in situ remediation and can be applied at contaminated sites on a large scale. Especially when combining the remediation potential with other cost-effective technologies, such as the use of the biomass for biofuel production, the phytoremediation technique holds great potential as an environmental cleanup technology.

Metal ions in sewage sludge are mainly produced by the industrial world

Sewage sludge is an important fertilizer but application also may result in contamination of soil by Cu, As, Pb, Zn, and Hg and others.

Many soils in the industrial world are contaminated with toxic metals which cause health hazards and disrupt life cycles of flora and damage whole ecosystems. Toxic effects can arise from natural processes and from aerial and aquatic sources. It is accepted that toxic metals, metalloids and their toxicity are largely affected by the physico-chemical nature of the soil environment which leads to a decreased or increased mobility. Clay minerals e.g., montmorillonite and kaolinite, and iron and manganese oxides are important determinants of metal availability in soils.

Although elevated levels of toxic heavy metals occur in some locations e. g. in volcanic soils, hot springs, and deep sea vents, average natural abundances are generally low. However ore mining and processing have disrupted natural biogeochemical cycles and can cause increased atmospheric release as well as deposition into aquatic and terrestrial environments. Major sources of pollution include the combustion of fossil fuels, mineral mining, and other industrial effluents and sludges, brewery and distillery wastes, biocides and preservatives including organometallic compounds.

Heavy metal ions with specific high toxicities

Cadmium

Cadmium (Cd) is a non essential heavy metal. It is used as protection against corrosion of steel and iron and widely applied in the electro industry (in accumulators and batteries), as ingredient in dyestuffs, and in the preparation of plastics (worldwide production of about 20,000 t/a). Due to the toxicity of cadmium its use in these applications becomes more and more reduced.

The main target organs for cadmium are kidney and liver with critical effects occurring when a content of 200 μg g^{-1} (wet weight) is reached in the kidney cortex.

The present intake of Cd suggested by FAO/WHO (1992) as tolerable limit amounts to 7 μg Cd kg^{-1} per kg body weight and a daily intake of 1 μg kg^{-1} body weight d^{-1}. For a person with 70 kg weight, a maximum of

70 µg d^{-1} Cd should not be surpassed, but the recommended dose is only half, i.e., 35 µg d^{-1} (Enquete Kommission, 1994). The closeness between actual intake and suggested maximum is one of the reasons for the concern about the Cd levels in soils and water which can cause an inhibition of microbial soil respiration and other microbial processes already at <1mg kg^{-1} soil and a reduced growth depression of 1.8 mg kg^{-1} soil.

Cd reaches the environment from a variety of sources and concern results from the contamination of soils by application of phosphate fertilizers (Schütze et al., 2003).

The mean input of Cd by mineral and further fertilizer application, and by deposition from the atmosphere is about 7,9 g Cd ha^{-1} a^{-1}, while the removal by plant uptake and leaching and erosion amounts to about 1,0 g Cg ha^{-1} a^{-1}.

The average Cd input by P-fertilizer supply for the average crop requirement amounts to about 7 g ha^{-1}a^{-1}. For the Cd pathway soil → plant → foodstuff → man uptake and resorption can be calculated. The resulting total Cd burden for man was already ca. 40% of the accepted load to approach the "no observed adverse effect levels" (NOAEL).

The mean Cd-entry from inorganic P-fertilizer with an average Cd-content of 90 mg kg^{-1} P$_2$O$_5$ average 5.6 g Cd ha^{-1}a^{-1}, while that from application of organic ferilizers varies between 0.01 and 2.12 g ha^{-1}a^{-1} (Table 7.3).

The values of soil Cd-entry by P-fertilizer indicates that in 85% of the agricultural land the Cd contents will increase. Only if the Cd-contents in P-fertilizers average less than 0.8 g ha^{-1}a^{-1} the present Cd-contents in agricultural soils can be diminished (Schütze et al., 2003).

Cd-content of soils is mostly expressed in total contents of Cd extracted with *aqua regia* with consideration of specific soil conditions (Throl, 2000; Ingwersen et al., 2000). The plant effective Cd content is expressed by the Cd amount extractable with aqueous ammonium nitrate solution (Prüess, 1992).

Table 7.3: Mean Cd -amounts in various P-fertilizers, related to P$_2$O$_5$– contents (Schütze, 2003).

Commercial fertilizer	Unit	Cd content
Triple phosphate	mg (kg P$_2$O$_5$)$^{-1}$ Cd	62
Diammonium phosphate	mg (kg P$_2$O$_5$)$^{-1}$ Cd	61
Potassium-ammonium phosphate	mg (kg N)$^{-1}$ Cd	1,4
Korn-Kali	mg (kg K$_2$O)$^{-1}$ Cd	0,23
Carbonate of lime	mg (kg Ca)$^{-1}$ Cd	1,04

Mercury

Mercury is similarly as Cd a non essential heavy metal. It is a liquid at room temperature with a density 14.5 higher than water. It has a high volatility and air in equilibrium with liquid mercury contains 14 mg Hg m^{-3} at 20° C. The maximum allowable non toxic level of Hg in air amounts to 0.05 mg Hg m^{-3} air.

Because of its volatility Hg evaporates in the surrounding air; this can be used to trace the presence of other metals in ores, since it is often combined with lead, zinc ores or other metals such as silver.

Worldwide production of Hg amounts presently to about 3´400 tons per year and most of it is used for electrolysis of NaCl to produce Cl$_2$-gas and NaOH. Losses of Hg in this process average 150–250 g Hg per t Cl$_2$. Due to the toxicity of mercury a worldwide ban of this metal is currently debated.

Differences between the global cycle of Hg compared to other toxic trace metals are microbial methylation of Hg^{2+} ions to mono- and dimethyl-mercury species.

$$Hg^{2-} \qquad \rightarrow \qquad CH_3Hg^{\cdot} \qquad \rightarrow \qquad (CH_3)_2Hg$$
$$\text{soluble in water} \qquad\qquad \text{insoluble in water, volatile}$$

Monomethylmercury is water soluble and is the major mercury species found in fish. Dimethyl mercury is insoluble in water, volatile and less toxic than methylmercury (Fig. 7.3).

Accidents in connection with Hg originate mainly from methyl-Hg because of its toxicity for the nervous system. Since it even surpasses the barrier of the placenta it also damages new born childs. This happened in Japan to people with high fish consumption from contaminated areas and in countries where corn was treated with Hg-phenyl containing "fungizides for seed treatment" and also in countries such as Irak, Pakistan and Guatemala.

Biomethylation means a mobilization or remobilization of mercury resulting in water soluble toxic mercury compounds in milk, meat and eggs. The target organ of methylmercury in humans is the brain upsetting the metabolism of the nervous system. In addition the main toxic effects of inorganic mercury aim at the kidney and liver.

In the poisoning case in Minamata (Japan), infants got serious brain damages, but were born to mothers who showed slight or no symptoms of mercury poisoning. In consequence the WHO suggested that the intake for adults should be less than 0.3 mg of total mercury and not more 0.2 mg methylmercury per person and week.

There are sensitive methods available to detect mercury at levels in the µg kg^{-1} range, but the high volatility of many mercury species causes difficulties in analysis.

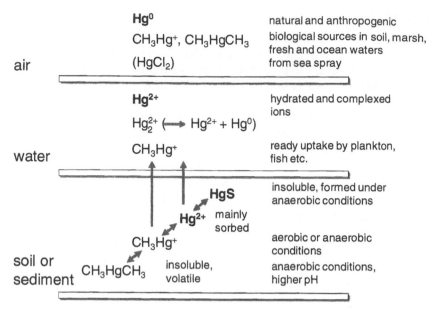

Fig.7.3: Mercury compounds in air, water and sediments and soil.

Lead

This metal is commonly used in accumulators, batteries, in water tubes, roof sealing, and formerly as addition to gasoline as anti-nocking agent to allow a higher compression of the gasoline-O_2-mixture in the motor and thereby an increased use of the energy of gasoline. The worldwide production presently is about 3.1 Mio. tons per year.

Large amounts of the lead in the air is derived from leaded gasoline in countries where this fuel additive is still used. Lead in gasoline damages the catalysts used for cleaning the exhaust of automobiles and trucks. From the leaded gasoline 75% of the lead is evaporated into the atmosphere as fine particle with diameters less than 2 µm.

Environmental legislation policy has resulted in increased sales of lead free petrol in the western world, therefore the amount of lead being released into the atmosphere is declining.

The average anthropogenic emission rate in the latter half of the nineteenth century was 22×10^6 kg a^{-1} due mainly to the smelting of lead ores and burning of coal.

About 94% of lead in the atmosphere results from anthropogenic sources. In cities with high density of car traffic and combustion of coal the lead contents in the atmosphere may be even higher. Still in low income countries leaded gasoline is used.

Lead contents in water tubes with a pH below 7 is very low since surfaces have a protective layer of Ca- and Mg-carbonates which inhibit the dissolution of lead. In areas with water below pH 5 lead is relative soluble and can contain up to 1 mg Pb L^{-1}.

Lead accumulates in the body and around 90% is transferred into bones, but can also be released again. The world health organisation limits the maximum intake to 3 mg Pb per week and person, for children to less than 1mg per week (Fig. 7.4).

The absorbed lead arrives mainly in the blood and remains to 90% in the red blood cells with an average residence time of at least one month. However, this remaining time depends strongly on the previous exposition to lead. Average lead contents in blood amount to 10–20 µg Pb 100 ml^{-1}. 25–40% remain in the soft body parts and 15% in the bones. The average residence time amounts to about one month with strong deposits in the hair and lungs. Lead in hair is a strong indicator of exposition in previous months.

Lead causes damages in blood formation and functions of brain and several biochemical metabolic reactions, since lead is tightly bound to a multiplicity of molecules such as amino acids and proteins, e.g., haemoglobin and many enzymes, RNA and DNA, and thus disrupts many metabolic pathways. Poisoning with lead causes high blood pressure; lead

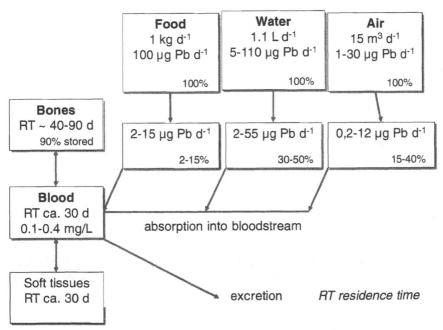

Fig. 7.4: The daily intake of lead for grown up persons and distribution in the body.

contamination of soils can cause lead contamination of fruits, crops and nutrients. However, the soil plant transfer factor of lead is relatively low (Table 7.1). Lead is particular dangerous for children and younger people.

Transformation of metal ions in soils

In soil are many organisms (bacteria, fungi, yeasts) capable to alkylate heavy metals (Table 7.4) with methylcobalamin as a potential methylgroup donor. The formed species differ in terms of toxicology and environmental fate. Microbial methylation plays important roles in the biogeochemical cycling of these metals and possibly in their detoxification (Bentley, 2002).

Table 7.4: Organisms involved in the alkylation of heavy metals (indicated by x).

	As	Hg	Sn	Se	Pb
aerobic					
Fungi					
Aspergillus	x	x		x	
Candida	x				
Gleocladium	x				
Neurospora	x	x			
Penicillium	x			x	
Saccaromyces	x	x			
Schizophyllum				x	
Scopulariopsis	x	x		x	
Trychophyton	x				
Bacteria					
Aeromonas	x			x	x
Acinetobacter					x
Alkaligenes					x
Bacillus		x			
Eschericchia	x	x			
Flavobacterium	x			x	x
Klebsiella		x			
Mycobacterium		x			
Pseudomonas	x	x	x	x	
anaerobic					
Clostridium		x			
Desulfovibrio		x	x		
Methanobacterium	x				

Under certain conditions, microbial metal reduction can mobilise toxic metals with potentially calamitous effects on environmental health; microorganisms are known to catalyze the reduction of Fe^{3+}, Mn^{4+}, Cr^{6+}, Hg^{2+}, Co^{3+}, Pd^{2+}, Au^{3+}, Ag^+, Mo^{6+}, V^{5+}, As^{5+}, Se^{6+} and radionuclides including U^{6+}, Np^{5+} and Tc^{7+} (Lloyd, 2003) (Table 7.5).

Table 7.5: Organisms involved in the reduction of heavy metal elements°.

Element	Organisms	Comment
Cr	Many spp. of *Pseudomonas*, *Aeromonas*, *Clostridium*, *Citrobacter*, *Bacillus*, *Streptomyces* and *Desulfovibrio*	Extensive contaminations with industrial Cr. Reduction of highly toxic and mobile Cr^{6-} to less toxic Cr^{3-} (e.g. $Cr(OH_3)$. Reduction by aerobes with glucose and other organics as electron donor. Desulfovibrio uses H_2 as donor
U	*Geobacter metallireducens*, *Shewanella putrefaciens*, *Desulvofibrio spp.*	Reduction of highly soluble U^{6-} to highly insoluble U^{4-} (e.g. UO_2)
Se	Many spp. of *Pseudomonas*, *Flavobacterium*, *Citrobacter*, *Clostridium*, *Thiobacillus* and others	Aerobes can reduce Se^{6-} to Se^{4-}, using organic acids or sugars, or to selenide (Se^{2-}).
Mo	*Pseudomonas guillermondi*. *Sulfobolus acidocaldrius* and others	Mechanisms of Mo^{6-} to Mo^{5-} is accomplished biotically
Mn	*Leptothrix*, *Crenothrix*, *Sphaerotilus*	Most organisms that reduce Fe^{3-} also reduce Mn^{4-}

° Data compiled by Lovely, 1993.

REFERENCES

Alloway WH, 1968: Agronomic controls over environmental cycling of trace elements. Adv Agron. 20, 235–74.

Alloway BJ, ed., 1994: Heavy metals in soil. Springer, pp 384.

Bentley R, Chasteen TG, 2002: Microbial methylation of metalloids: arsenic, antimony, and bismuth. Microbiol. Mol Biol. Rev. 66, 250–71.

Bode HP, Dumschut M, Garotti S, Fuhrmann GF, 1995: Iron sequestration by the yeast vacuole, Eur. J. Biochem.228, 397–342.

Burgstaller and Schinner, 1993: Leaching of metals with fungi. J. Biotechnol. 27, 91–116.

Chander K, Dyckmans J, Joergensen RG, Meyer B, Raubuch M, 2001. Different sources of heavy metals and their long-term effects on soil microbial properties. Biol. Fertil. Soils 34, 241–247.

DeVries, Bakker DJ, Sverdrup HU, 1998: Manual for calculating critical loads of heavy metals for aquatic ecosystems, Report 165, DLO, Winand Staring Centre; Wageningen, The Netherlands.

Doelman P, Jansen E, Michels M. van Til M, 1994: Effects of heavy metals in soil on microbial diversity and activity as shown by the sensitivity resistance index, an ecologically relevant parameter. Biol. Fertil. Soil 17, 177–184.

Doelman P, Hanstra I, 1979: Effects of lead on the soil bacterial microflora. Soil Biol. Biochem.11: 487–491.

Enquete Kommission, 1994: Deutscher Bundestags: Verantwortung für die Zukunft. Vol. 5, Bonn, 213 pp.

Evangelou M, Ebel M, Schäffer A, 2007: Chelate assisted phytoextraction of heavy metals from soil. Effect, mechanism, toxicity, and fate of chelating agents. Chemosph. 68, 989–1003.

Gadd GM, Lawrence QS, 1996: Demonstration of high-affinity Mn^{2-} uptake in Saccharomyces cerevisiae—specifity and kinetics. Microbiol.142, 1159–1167.

Giller K, Witter E, Mc Grath SP. 1998: Toxicity of heavy metals to microorganisms and microbial processes in agricultural soils: a review: Soil Biol. Biochem. 30, 1389–1414.

Ingwersen J, Streck T, Utermann J, Richter J, 2000: Ground water preservation by soil protection: Determination of tolerable total Cd contents and Cd breakthrough times. J. Plant Nutr. Soil Sci. 163, 31–40.

Kägi JHR, Schäffer A, 1988: Biochemistry of metallothionein. Biochem. 27, 8509–8515.

Klees C, 1993: Untersuchungen über den Einfluss organischer Säuren auf das Adsorptions- un Desorptionsverhalten von Phosphat an Goethit. Diss. TU-Braunschweig, Naturwiss. Fak., 167 pp.

Knoche H, Brandt P, Viereck-Götte L, Böken H, 1999: Schwermetalltransfer Boden-Pflanze: Umweltbundesamt 11.

Kuperman RG, Carreiro MM, 1997: Soil heavy metal concentrations. Microbial biomass and enzyme activities in a contaminated grassland ecosystem. Soil Biol. Biochem. 29, 179–190.

Lloyd JR, 2003: Microbial reduction of metalloids. FEMS Microbiol. Rev. 27, 411–425.

Loveley DR, 2000: Environmental microbe-metal interactions. ASM Press, Washington DC.

Messerle BA, Schäffer A, Vasak M, Kägi JHR, Wüthrich K, 1992: Comparison of the solution conformations of human $[Zn_7]$-metallothionein-2 and $[Cd_7]$-metallothionein-2 using nuclear magnetic resonance spectroscopy. J. Mol. Biol. 225, 433–443.

O'Neill P, 1994: Environmental chemistry. Second edition, Chapman Hall, pp 268.

Prüess A, 1992: Vorsorgewerte und Prüfwerte für mobilisierbare potentiell ökotoxische Spurenelemente in Böden. Wendlingen, Grauer, 145 pp. ISBN 3-98033063-4-8.

Raskin I, Smith RD, Salt DE, 1997: Phytoremediation of metals: using plants to remove pollutants from the environment. Curr. Opin. Biotechnol. 8, 221–226.

Rawlings DE, 1997: Mesophilic, autotrophic bioleaching bacteria, description, physiology and role. In: biomining; theory, microbes and industrial process. Rawlings DE,ed., Springer, Berlin. p 229–245.

Reber HH, 1992: Simultaneous estimates of the diversity and the degradative capability of hevy-metal affected soil bacterial communities.Biol. Fertil. Soils 13, 181–186.

Roane TM, 1999: Lead Resistance in Two Bacterial Isolates from Heavy Retal–Contaminated Soils. Microb. Ecol. 37, 218–224.

Ryan RP, Ryan DJ, Dowling DN, 2005: Multiple metal resistant transferable phenotypes in bacteria as indicators of soil contamination with heavy metals. J. Soils Sed. 5, 95–100.

Schippers A, Sand W, 1999: Bacterial leaching of metal sulfides procedes by two indirect mechanisms via thiosulfate or via polysulfides and sulphur. Appl. Environ. Microbiol. 65, 319–321.

Schütze G, Becker U, Dämmgen H-D, Nagel A, Schlutow A, Weigel H-J, 2003: Risikoabschätzung der Cadmium-Belastung für Mensch und Umwelt infolge der Anwendung von cadmiumhaltigen Düngemitteln. Landbauforschung Völkenrode 53: 63–170.

Starkey RL, Waksman SA, 1943: Fungi tolerant to extreme acidity and high concentrations of copper sulphate. J. Bact. 45, 509–519.

Throl C, 2000: Derivatization of ecotoxicologically based soil quality data. Proc. Altlasten in Böden UN/ECE. Fraunhofer Inst., Schmallenberg, 133 pp.

Uchida S, Tagami K, Hirai I, 2007: Soil-to-plant transfer factors of stable elements and naturally occurring radionuclides: (1) Upland Field Crops Collected in Japan J. Nucl. Sci. Technol. 44, 628–640. (2) Rice Collected in Japan. J. Nucl. Sci. Technol. 44, 779–790.

INDEX